Palgrave Historical Studies in the Criminal Corpse and its Afterlife

Series Editors
Owen Davies
School of Humanities
University of Hertfordshire
Hatfield, UK

Elizabeth T. Hurren
School of Historical Studies
University of Leicester
Leicester, UK

Sarah Tarlow
History and Archaeology
University of Leicester
Leicester, UK

Aim of the Series:
This limited, finite series is based on the substantive outputs from a major, multi-disciplinary research project funded by the Wellcome Trust, investigating the meanings, treatment, and uses of the criminal corpse in Britain. It is a vehicle for methodological and substantive advances in approaches to the wider history of the body. Focussing on the period between the late seventeenth and the mid-nineteenth centuries as a crucial period in the formation and transformation of beliefs about the body, the series explores how the criminal body had a prominent presence in popular culture as well as science, civic life and medico-legal activity. It is historically significant as the site of overlapping and sometimes contradictory understandings between scientific anatomy, criminal justice, popular medicine, and social geography.

More information about this series at
http://www.springer.com/series/14694

Shane McCorristine
Editor

Interdisciplinary Perspectives on Mortality and its Timings

When is Death?

Editor
Shane McCorristine
Independent Researcher
Cambridge, UK

Palgrave Historical Studies in the Criminal Corpse and its Afterlife
ISBN 978-1-137-58327-7   ISBN 978-1-137-58328-4 (eBook)
DOI 10.1057/978-1-137-58328-4

Library of Congress Control Number: 2017944547

© The Editor(s) (if applicable) and The Author(s) 2017. This book is an open access publication.
The author(s) has/have asserted their right(s) to be identified as the author(s) of this work in accordance with the Copyright, Designs and Patents Act 1988.
Open Access This book is licensed under the terms of the Creative Commons Attribution 4.0 International License (http://creativecommons.org/licenses/by/4.0/), which permits use, sharing, adaptation, distribution and reproduction in any medium or format, as long as you give appropriate credit to the original author(s) and the source, provide a link to the Creative Commons license and indicate if changes were made.
The images or other third party material in this book are included in the book's Creative Commons license, unless indicated otherwise in a credit line to the material. If material is not included in the book's Creative Commons license and your intended use is not permitted by statutory regulation or exceeds the permitted use, you will need to obtain permission directly from the copyright holder.
The use of general descriptive names, registered names, trademarks, service marks, etc. in this publication does not imply, even in the absence of a specific statement, that such names are exempt from the relevant protective laws and regulations and therefore free for general use.
The publisher, the authors and the editors are safe to assume that the advice and information in this book are believed to be true and accurate at the date of publication. Neither the publisher nor the authors or the editors give a warranty, express or implied, with respect to the material contained herein or for any errors or omissions that may have been made. The publisher remains neutral with regard to jurisdictional claims in published maps and institutional affiliations.

Cover illustration: © David Anthony/Alamy Stock Photo; all rights reserved, used with permission

Printed on acid-free paper

This Palgrave Macmillan imprint is published by Springer Nature
The registered company is Macmillan Publishers Ltd.
The registered company address is: The Campus, 4 Crinan Street, London, N1 9XW, United Kingdom

Acknowledgements

This book emerged from the "Harnessing the Power of the Criminal Corpse" project at the University of Leicester. Generously funded and made open access by the Wellcome Trust (Grant number: 095904/Z/11/Z), this project brought together a team of interdisciplinary scholars to explore themes relating to crime, punishment, execution, magic, and the notorious dead. I would like to thank: all the members of the team for their collegiality and curiosity over several years working together; our Principal Investigator, Sarah Tarlow, who co-organised the 2015 conference at Leicester, "When is Death?", that led to this book; the contributors who shared their wonderful research with us in this book; and Emily, Carmel, and everyone at Palgrave Macmillan, who were patient and helpful throughout the process.

Contents

1 **Introduction**   1
Shane McCorristine
When is Death?   1
The Traditional Irish Wake   4
Chapters in This Book   11
Notes   14

2 **Being Dead in Shakespearean Tragedy**   17
Mary Ann Lund
The Borderlands of Death: Renaissance Tragedy   19
Becoming Dead in Shakespeare   21
Textual Deaths in Hamlet   24
Produce the Bodies: King Lear   27
Notes   30

3 **"A Candidate for Immortality": Martyrdom, Memory, and the Marquis of Montrose**   33
Rachel Bennett
Montrose: Covenanter and Cavalier   35
The Execution of Montrose   36
"An Honourable Reparation"   39

	"A Most Convincing Form of Testimony"	40
	Conclusion	43
	Notes	45

4	Overcoming Death: Conserving the Body in Nineteenth-Century Belgium	49
	Veronique Deblon and Kaat Wils	
	Conserving the Dead Body: The Preparations of Frederik Ruysch	51
	Death in Nineteenth-Century Belgium	53
	Between Life and Death: The Preparations of Adolphe Burggraeve	55
	Embalmment Mania: Jean-Nicholas Gannal's Conservation Method	58
	The Application of Conserving Procedures	60
	Overcoming Death: The Post-Mortem Subject	61
	Conclusion	63
	Notes	63

5	Premature Burial and the Undertakers	69
	Brian Parsons	
	The Changing Context of Disposal	71
	Premature Burial and the Undertakers	75
	Notes	82

6	The Death of Nazism? Investigating Hitler's Remains and Survival Rumours in Post-War Germany	87
	Caroline Sharples	
	Notes	99

7	Death's Impossible Date	103
	Douglas J. Davies	
	Introduction	103
	Chronological Precision: Cultural Measures of Death	103
	Life-Course Narratives	108
	Existential Anticipation	113
	Conclusion	114
	Notes	115

8	**The Legal Definition of Death and the Right to Life** Elizabeth Wicks	119
	Introduction	119
	The Legal Definition of Death	120
	Higher-Brain Death: Its Proponents and Problems	122
	Death and the Right to Life	124
	Life, Death, and Embodied Selves	126
	Conclusion	129
	Notes	130
9	**The Last Moment** Jonathan Rée	133
	Notes	142
10	**Afterword** Thomas W. Laqueur	145
	Notes	152
Further Reading		155
Index		163

EDITOR AND CONTRIBUTORS

About the Editor

Shane McCorristine is an interdisciplinary historian with research interests in the history of Arctic exploration, the cultural geography of extinction, and the place of the supernatural in the modern world. Between 2013 and 2015 he was a Wellcome Postdoctoral Fellow on the "Harnessing the Power of the Criminal Corpse" project at the University of Leicester. His most recent book is *William Corder and the Red Barn Murder: Journeys of the Criminal Body* (Palgrave Macmillan, 2014).

Contributors

Rachel Bennett University of Warwick, UK

Douglas J. Davies Durham University, UK

Veronique Deblon University of Leuven, Belgium

Thomas W. Laqueur University of California, Berkeley, USA

Mary Ann Lund University of Leicester, UK

Brian Parsons London, UK

Jonathan Rée London, UK

Caroline Sharples University of Roehampton, UK

Kaat Wils University of Leuven, Belgium

Elizabeth Wicks University of Leicester, UK

LIST OF FIGURES

Fig. 4.1   Anatomical preparation of a newborn child
            (Museum for the History of Medicine, Ghent.
            Collection: University Museum, Ghent)   57
Fig. 4.2   Anatomical preparation of a girl with hand
            (Department of Basic Medical Sciences, Anatomy
            and Embryology Research Group. Collection:
            University Museum, Ghent. Photographer:
            Benn Deceuninck)   58

LIST OF TABLES

Table 7.1 Experienced presence of the dead   110
Table 7.2 Relation—experience   111

CHAPTER 1

Introduction

Shane McCorristine

WHEN IS DEATH?

When is a person dead? This is a difficult and perhaps impossible question to answer—but it is fascinating to think about and explore. It is a question that has many different meanings and iterations depending on the context in which it is asked. Today, if we ask a physician to calculate when the death of someone has occurred or will occur, we will get approximate times based on particular biological indications drawn from their training and technologies. If we ask a bereaved family the same question we will probably get very different answers based around other, non-biological, indicators, customs, and timeframes.[1] These might include the times when the person lost their identity, when the ritual of burial or cremation took place, or when legal or criminal proceedings decided on death certification, or the status of a present, but "brain dead" person. Asking the same question 200 years ago, of course, would have resulted in a different set of answers altogether.

Dwelling on the question of death timings, then, is like peeling back the layers of an onion—it soon starts to open up a range of further temporal issues and uncertainties about when life ends and death begins.

S. McCorristine (✉)
Independent Researcher, Cambridge, UK
e-mail: shanemccorristine@yahoo.ie

© The Author(s) 2017
S. McCorristine (ed.), *Interdisciplinary Perspectives on Mortality and its Timings*, Palgrave Historical Studies in the Criminal Corpse and its Afterlife, DOI 10.1057/978-1-137-58328-4_1

When does a person die—really, fully, and absolutely? Is it when signs of life cannot be found in a body? Is it when a person, corpse, or life-force no longer has any agency in the world? Or is it when living people recognise a person as socially dead and outside the frame of everyday relationships? Death as an event is clearly something that comes with multiple temporalities for living and dying people. These death timings cut across biological, social, legal, and ethical ways of defining when a person is and when a person is no longer. How, then, do we begin to investigate the stages, sequences, and chronologies that flow through the time between the end of life and the start of death?

This book arises from a conference entitled "When is Death?" that was hosted by the University of Leicester in 2015. This gathering was part of the work of an experimental interdisciplinary team of researchers at Leicester who were funded by the Wellcome Trust to investigate the power of the criminal corpse in European culture and history.[2] The criminal corpse was chosen as an object of focus because it was a force that harnessed legal, medical, and popular discourses about death and agency in the period 1500–1900. Individually, we researched topics such as the technology of the gibbet, criminal execution and dissection, folk-beliefs about the dead, and bioethics.[3] Collectively, we came to realise that we all had concerns about the "when" of the deaths we were investigating. These included: living individuals who became socially dead; criminals who survived their own executions; and deaths whose timelines were stretched out for decades by the unending spectacles of post-mortem punishment. By looking at the journeys of particular bodies towards and beyond death we argued that the timing of death—something that at first seemed so certain and absolute—was deeply uncertain and open to interpretation. This raised a set of issues and contexts that called for an exploration of the question from multiple perspectives—an interdisciplinary endeavour which led to the contributions in this book.

Scholars working in the field of death studies regularly analyse and discuss the ideas of philosophers and thinkers as varied as Thomas Browne, Fustel de Coulanges, Martin Heidegger, Philippe Ariès, and Elisabeth Kübler-Ross. Yet ideas and themes in popular culture provide us with other and equally fruitful routes into thinking about death timings. I suggest that *Weekend at Bernie's* (1989)—one of the more unusual movies of the 1980s—offers us a range of ways to think about the timing of death. The plot of the movie can be summed up as follows:

during a baking hot New York summer Larry and Richard, two employees in an insurance corporation, uncover fraudulent accounting and go to their boss, Bernie Lomax, with the information. Realising that his crimes have been rumbled, Bernie hires a mobster to kill the two young men at a party he will host at his exclusive Long Island beach-house. However, instead the mobster has Bernie killed for having an affair with his wife. Excited about their weekend of partying at Bernie's house, Larry and Richard arrive to find Bernie dead in his chair. Thus begins a black comedic narrative in which Bernie's corpse retains its liveliness in many biological and social ways. Fearing that they will be suspected of Bernie's murder, Larry and Richard delay informing the police. As Bernie's house starts to fill up with partygoers they see that nobody realises that their quiet host, sitting on the couch in a pair of sunglasses, is actually dead. Rather, Bernie's friends greet and kiss him, massage him, haggle with him, and treat him as a living person. Larry and Richard therefore decide to "postpone" Bernie's death so that they might have a fun weekend. They respond to Bernie's social liveliness by acting as his guardian, explaining to the others that he is drunk. Without ever being noticed, they carry him, move his limbs about, and speak for him when the need arises.

It is quite remarkable how many social interactions the corpse has in the movie. Over the course of the weekend Bernie smokes, waves, and goes waterskiing; he is "murdered" a further two times by the mobster assassin (who comes to question his own sanity when Bernie won't die); and even has sex with his mistress who visits the beach-house to find out why he had not been in touch. This prompts the immortal line: "This guy gets laid more dead than I do alive!". Even more remarkable is the fact that, despite the sweltering weather on Long Island, Bernie's body does not decompose nor show any signs of rigor mortis.

Weekend at Bernie's demonstrates how forms of life and liveliness can continue beyond biological death if the person is still a part of social life. In this fantasy of death denial, the "when" of Bernie's death is not recognised by his friends and is pushed into a distant temporal horizon by the two young men so that they can enjoy themselves in his company. In an even more macabre turn of events, Bernie's corpse is reanimated—really this time—through voodoo magic in the sequel, *Weekend at Bernie's II* (1993). In other words, corpses have power and social lives through their own actions/reactions and the actions/reactions or assumptions of living people.

The Traditional Irish Wake

In its celebration of life and social interaction amid death, the plot of *Weekend at Bernie's* has much in common with the ritual of the traditional wake in Ireland. This funeral custom was a prominent feature of early modern and modern Catholic Irish culture and it still occurs today, albeit in a different form. In its earlier form, the wake was a special period of time marked out from the normal course of events by the watching over of the corpse before its burial, usually for two nights. Wakes in Ireland involved a ritual of sequences and timings that exposes some of the paradoxes surrounding death. On the one hand, the corpse became a time-telling object, displaying all the biological indications of death, but on the other hand the corpse continued to be regarded as socially alive and an active participant in the merriment of its send-off. We can therefore think of the wake as a ritual time during which the person and the corpse slowly transitioned to other states. The living marked these journeys by "resetting" the time of death through these customs.

We can trace out three temporal stages in the wake, each with their own crossing-cutting timelines: The first stage came in the lead-up to the wake; before death has arrived, it already has a temporal horizon. There was a general desire for death to be "done well" and it was known for the poor to endure privations in order to scrape together the means for a "fine wake" and a "decent funeral".[4] Old and sick people were seen as most likely to die in the short-term and this held out the tantalising possibility of a merry wake for the young people in the village. One commentator told the story of visiting a man who was ill with consumption, but expected to live, and found the kitchen full of men and women dressed in their Sunday best. Asking them why they were there, the man received the answer: "'We are waiting for the wake'. I inquired who was dead, 'No one; but the man within is all but dead and we are chatting a bit that we may help the widow to lift him when the breath goes out of his body'".[5]

Another nineteenth-century folklore collector reported:

> So great was the amusement carried on at an Irish wake-house, that all the persons of both sexes were anxiously on the look-out for the deaths of certain old men and women in the parish. When some of the young men met a very old poor woman, the usual salutation was: 'How are you to-day, Biddy? you are living a long time. What time will you give us the pleasant night over you? We are expecting it now for the last seven years, and you are still as tough as ever, though you are near a hundred years old!'[6]

Irish folklore is full of examples of signs that a death was impending: when a crow lands on ones shoulder; when a frog enters a house; hearing a crying or knocking at the door. If the corpse at a wake was not stiff, people interpreted it as a sign that another person would die soon enough. Omens observed on the funeral walk also presaged a new cycle of death: when a gap developed in the cortege it was said that there would be another death in the village. This folklore echoed vernacular senses of time as being fragmented and contingent—futures could be known and, to some extent at least, predicted and controlled. When a death was sudden, untimely, or was that of a child, it is notable that wakes were more private and mournful than the traditional merry wake.

Through customs relating to clocks and watches, folklore also shows us that people literalised the idea that death was timed by stopping the household clock when a person died. This is a story collected on Rathlin Island:

> [Informant] 1: The clocks always stopped [at the time of a death].
>
> [Informant] 2: Well, I had never heard that story about clocks or anything else to do with clocks. And there was a man here died…he was very ill, and he died this night…And here, Francis had not long got that clock and it kept excellent time, only had it a few weeks. And for some reason, what time…did he die? And here the clock had stopped at that exact time. Well…I thought that was a bit funny, the clock should stop then you know. I was standing, the day of the funeral and he'd to go by that way you see, to go up to the chapel, the body was to go up there and Francis was away at it, and I was standing here…I was watching them going by up there, you know, following the funeral. Honest, I'll never forget it, and the clock all of a sudden, just as the body went by started 'tick tock tick tock', it started up again. It scared me stupid. I says, 'Why should it start up again then?' You know, this was about two days later, the clock started. It was the weirdest thing.
>
> Inf. 1: You see, it was always the custom, they always stopped the clock whenever someone died in the house, they used to stop it as soon as they died.
>
> Inf. 2: So you wouldn't go into the house and ask them.
>
> Inf. 1: Aye, that was the reason, that, when you went in, you just looked at the clock and then you knew what time they died at.[7]

Old people used to say that if the cock crowed at midnight it was a sign of impending death, or that strange sounds in the clock presaged death.[8] It was also believed that the "dead man's tick", a sound in the wall like the ticking of a clock, indicated a death was imminent.[9] The wake, then, was the ritual time after the clock—or the passing of normal time—was temporarily stopped.

The second stage in the wake commenced after the moment of death was noted. However, this moment was dangerous and not easy to time. In Connemara, County Galway, when a man was dying of consumption it was customary to tie some unsalted butter in a piece of cloth and hang it up in the rafters. Just as the sick person gave his last breath, the consumption left the body and looked to enter another. If there was a relative present it would enter them, but if not it would go up into the butter, which was then taken down and buried. The relatives, meanwhile, stayed outside the house "til he's dead—and wel dead".[10]

When a person died in Ireland, news spread quickly. Work in the fields ceased and preparations were made to care for the corpse and stock the house (or barn) for the wake with funeral provisions (pipes, tobacco, alcohol, etc.). In the meantime, the corpse was laid out and prepared by female mourners. The body was first allowed some undisturbed time so that the soul could communicate with God before the women washed the corpse, wrapped it in a sheet, habit, or suit, and then placed it on its back on a table, door, or bed. In many cases a crucifix was placed on the breast and rosary beads were entwined in the fingers. It was important that no tear was shed on the body during this stage (which took two to three hours) because it had not "settled" yet. Ritualised crying, or "keening" [Irish: *caoineadh*] would only begin after the women's preparations were done and they withdrew from the body. This stage of preparation and caring for the body also served to mark and pass over a time of physiological changes. The settling of the corpse was coincident with the periods of rigor mortis and algor mortis, during which muscles contract, the joints are immobilised, the body cools, and the skin loses its elasticity. In this context, caring for the corpse by positioning it peacefully, closing the eyes, and tying the jaw, also performed the crucial task of ascertaining that the person really was biologically dead.

The third stage of the wake was a time to ritually mark the bonds that bound the living and the dead. Like death rituals in other cultural contexts, the motivations for attending a wake in Ireland were varied: people gathered around the corpse in order to celebrate life and remember the dead; to guard the corpse from evil; to ensure that death had really occurred; and to placate the soul of the dead.[11] Wakes traditionally took place over two nights before burial on the third day, and the corpse was never to be left alone. This wake-time was passed, then, in a slow manner with candles always burning, or as often as the householder could afford. Wake-time was not simply passed in an unconscious manner—rather, people stayed awake through structured time. Time was periodically marked out by rituals and rhythms that reminded mourners about the presence of the corpse and the well-being of the dead. With seats arranged around the walls, encircling the body, the *bean chaointe* [keener] led periodic laments; spontaneous or planned rosaries were said; and the priest visited to lead prayers for the dead. This scene was described by an English traveller in the seventeenth century:

> [The mourners] spend most of the night in obscene stories, and bawdy songs, until the hour comes for the exercise of their devotions; then the priest calls on them to fall to their prayers for the soul of the dead, which they perform by the repetition of Aves and Paters on their beads and close the whole with a *de profundis* and then immediately to the story or song again, till another hour of prayer comes; thus is the whole night spent 'till day.[12]

These moments had the power to effect reconciliation between the dead person and his or her surviving friends and family and achieve their incorporation with the inhabitants of the afterlife.[13] There was, then, another temporal horizon gestured at in the prayers for the dead, for the amount of time that the soul of the deceased spent in purgatory concerned those who attended wakes. In 1813, a Purgatorian Society was founded in Dublin and, for the price of a penny a week, every subscribing member was entitled to have post-mortem masses said for them and their family to relieve the burden of time spent in purgatory. Members of the society also recited the Office of the Dead Latin at wakes.[14] The time being passed by the dead in the afterlife after burial was also marked by the recurring recital of prayers by the living as well as the custom of the "month's mind" requiem mass.

In contrast to these devotional means of timing and relating to death, unruly and boisterous customs were kept to constantly re-socialise the dead person as the body was passing further and further into death and decomposition. These customs—increasingly targeted by ecclesiastical authorities in the nineteenth century as uncivilised and superstitious—included feasting, alcohol consumption, wake games, and other licentious behaviours. Together, these customs symbolised the hospitality that the dead person, as host, was providing for the guests and the reciprocal celebration and sustenance that the living provided for the dead. The dual role of the corpse at the wake goes some way towards explaining why visiting observers and critics could be shocked by the swift transitions between "holy sorrow" and "orgies of unholy joy" at Irish wakes.[15] They were an enactment of community, with the merriment and mourning of each wake a prelude to another—an endless cycle of death and rebirth, of hosting and being hosted. As Maria Edgeworth put it in her *Castle Rackrent* (1800):

> Deal on, deal on, my merry men all,
> Deal on your cakes and your wine,
> For whatever is dealt at her funeral to-day
> Shall be dealth to-morrow at mine.[16]

Tobacco was a central feature of the merry wake, both as a stimulant with symbolic properties and as a means of passing wake-time. The plate of tobacco was therefore placed on a table over the corpse, or on the corpse itself, or underneath the table, and was offered on arrival to all guests by a young server as the "dealing" or gift of the dead. Tobacco was taken as snuff or smoked in a new clay pipe, and even if one was not a smoker, it was customary to have a few puffs in memory of the deceased and whisper *Beannacht Dé ar anamnacha na marbh* [Lord have mercy on the dead]. Tobacco was also provided at the graveyard and in the west of Ireland it was known for celebrants to leave their pipes on the grave as a token.[17] According to folklore collected in the twentieth century, the origin for the use of tobacco at wakes came from the time of Christ when the watchers over his tomb found it hard to stay awake. Suddenly a plant appeared and beside it was a pipe. One of the watchers plucked the leaves of the plant, put it in the pipe, and smoked it, and since then they are given at wakes.[18]

While the ritual use of tobacco and whiskey, or poitín, were still common features of wakes into the twentieth century, the practise of wake games declined swiftly in the period after the Great Famine (1845–1852) due to a fracturing of communities, steady decline in the use of the Irish language, and clerical criticisms of popular mortuary practices. These games were often aggressive and played by the young men who had the energy to stay awake all night. They included potato or turf-throwing in the dark, slapping games, wrestling matches, and other rough pranks. There were also many different kinds of mock trials against an unlucky booby, and cycles of "mobbing"—back and forth battles of wits among men that could lead to fights. The dead person symbolically participated in some of these games, as in "Lifting the Corpse", in which a stout man would lie prone on the floor with straight legs and four men tried to raise him off the floor with only one thumb each. The corpse was also a direct participant in rough play by being pushed around, or being made to hold a hand of cards or smoke a pipe. It was also a well-known trick to stitch a distracted mourners' coat-tails to the corpse's winding sheet.[19] The folkloric record is also rich with darkly humorous examples of the dead person "reviving" at the wake: this was typically achieved by the prankster tying a rope around the corpse before rigor mortis set in (especially if the deceased had a hunchback or bow legs) and then cutting it at an opportune moment to make the body sit up and frighten everyone present.[20]

To judge from edicts and directives issued by Catholic bishops stretching back to the seventeenth century, the Church was especially concerned about erotic and transgressive games such as "Building the Ship" and "Building the Fort". In the latter game, young men ran at each other with "spears" before one fell down as if mortally wounded:

> then all the hooded women came in again and keened over him, a male voice at intervals reciting his deeds, while the pipers played martial tunes. But on its being suggested that perhaps he was not dead at all, an herb doctor was sent for to look at him; and an aged man with a flowing white beard was led in, carrying a huge bundle of herbs. With these he performed sundry strange incantations, until finally the dead man sat up and was carried off the field by his comrades, with shouts of triumph.[21]

Clearly these games were linked to courtship rituals between young men and women, and indeed it was frequently said that more love matches were made at wakes than at weddings. This background motivation was brought to the foreground in the practice of mock marriages by mock priests at wakes. In an account of one of these marriages the priest, dressed "in robes of straw carrying rosary beads made of potatoes, surmounted by a frog for a cross" joined together two young people with the words: "'may the full blessing of the beggars descend upon you; may ye have plenty of ragged children'".[22]

The time spent with the dead was clearly considered necessary for a decent send-off for the deceased and for the community to engage in social interaction surrounding death and burial. However, on occasion, the length of time that wakes took before burial was a cause of concern for the authorities. This was chiefly due to the fact that the timeline of decomposition was not coincident with the social timeline that the body was still a part of. In 1871, for instance, an Irish labourer named Tehan died in Southwark, London, and was waked for some five days. A coroner's inquest report stated that since the man's death his large family had lived and slept with the corpse:

> Since the death of deceased a perpetual 'Irish wake' had been kept up, and every evening the friends of Tehan had met, drunk whiskey, and told tales of the dead, whilst women every now and again howled frightfully. The body was laid on the bed, and surrounded by eighteen candles. For three nights prayers had been said over it, and it had been 'sat by' with great ceremony. The face of the deceased was uncovered, and the body strewn with flowers.

The body "presented a shocking appearance" and was in an advanced state of decomposition. The coroner wanted to abolish the "disgraceful" practice and remove the body to a dead-house, but it was reported that the family resisted this move.[23]

It is a testament to the symbolic importance of merry wakes in Ireland that they continued to be held despite official directives from the Church. This suggests that the ritual activity of wake-time had a functional purpose for both the family affected and the wider community.[24] In gathering people together to mark out the deceased person's transition to a new state, the traditional Irish wake made death timely through rituals and customs. These customs began to fade in the second half of the nineteenth century as mortuary practices became "faster". Wakes were shortened to one night; games and keening were no longer held; and mourners were given less time to work out the paradox of death. As priests ensured that the corpse spent the second night in the Church (the "removal"), the funeral on the third day became the primary rite of passage. The care shown to the dead was reset into another, less socially interactive, timeframe: this is the situation that persists in Ireland today.

CHAPTERS IN THIS BOOK

Thomas Laqueur writes: "Death in culture takes time because it takes time for the rent in the social fabric to be rewoven and for the dead to do their work in creating, recreating, representing, or disrupting the social order of which they had been a part".[25] In their different ways, the contributors to this book address Laqueur's argument, suggesting that the time of death (if indeed there is one single point in time when life can be said to be extinct) is social, as much as it is biological.

In Chap. 2, "Being Dead in Shakespearean Tragedy", Mary Ann Lund looks at how speech acts indicate death timings on the English Renaissance stage. In his tragedies, William Shakespeare used the power of language to stage self-referential games of being dead and playing being dead. A scene in *Othello* provides a good example of this: When Othello first smothers Desdemona, he does not succeed in killing her: "Not dead? Not yet quite dead?". Othello continues smothering her until he states, "She's dead". However, even then, Desdemona's death is not certain at all. Death in Shakespeare, Lund argues, is a participatory process that is dependent on staging, performance, and language.

Moving from drama to political history, in Chap. 3, "'A Candidate for Immortality': Martyrdom, Memory, and the Marquis of Montrose", Rachel Bennett looks at the death and afterlife of James Graham, 1st Marquis of Montrose. Executed as a traitor in Edinburgh in 1650, Montrose was a key figure in the military and religious conflicts which ripped Britain apart in the mid-seventeenth century. However, as Bennett shows, far from being the final stage on Montrose's journey, his execution, or legal death, came some time after his excommunication and social death. This was followed by an exhumation and honourable reburial in 1661, while his scattered body parts were "re-membered" by royalists for centuries afterwards. The "when" of Montrose's death, Bennett suggests, was a matter of political debate and conflict.

Although for many of us in Europe and North America, the dead body is a distant and infrequent presence, spectacles like Gunther von Hagens's *Bodyworlds* exhibitions stage face-to-face encounters with conserved corpses for millions of ticket-holders. As Veronique Deblon and Kaat Wils show in Chap. 4, "Overcoming Death: Conserving the Body in Nineteenth-Century Belgium", these types of encounters, and the mixed emotions they provoke, have a long history. Deblon and Wils explore how new conserving procedures were developed in the nineteenth century in response to a growing disgust at the decaying corpse and a growing desire to create a corpse that looked as if it was sleeping. In Belgium, anatomists responded to this increasingly sentimentalised relationship with the dead by making anatomical preparations in a highly aesthetic manner. Although treated, injected, and embalmed, these prepared bodies and body parts seemed to convey the consoling message that death could be peaceful.

In contrast to this representation, many people were haunted by the disturbing message that some of the dead were perhaps not so "dead" after all. In Chap. 5, "Premature Burial and the Undertakers", Brian Parsons focuses his attention on how the fear of premature burial changed the way dead bodies were cared for and disposed of in nineteenth-century Britain. Far from being an irrational modern iteration of a primeval fear, concerns about premature burial arose due to a deficiency in the law which meant that physicians did not have to check for signs of life before certifying death. Because of this, people developed a range of strategies to ensure that they, or their loved ones, were truly dead. These included: delaying burial; safety coffins; paying for a cremation instead of a burial; or hiring an undertaker to embalm the corpse. In his survey of this landscape of disposal, Parsons concludes that by testing for death, undertakers gained "a new status as quasi-medical practitioners and helped shed the Dickensian image of disreputability inherited from their nineteenth-century forebears".

From Elvis Presley to Lord Lucan, many of the famous or notorious dead circulate as undead in popular culture, returning repeatedly in rumours, purported sightings, and conspiracy theories. Perhaps no figure illustrates this phenomenon more than Adolf Hitler, said to have committed suicide in his Berlin bunker in 1945 (but memorably imagined by *Monty Python's Flying Circus* as living out his days in a small guesthouse in Minehead, Somerset). In Chap. 6, "The Death of Nazism? Investigating Hitler's Remains and Survival Rumours in Post-War Germany", Caroline Sharples looks at the phenomenon of Hitler survival stories and traces their endurance to the failure of the Allies to conclusively identify his remains in 1945. As she persuasively argues, both the Allies and the Nazis before them cast doubt on the timing of Hitler's death in order to further their own interests. When put together with, on the one hand, obfuscation from the Soviet authorities who forensically examined the scene and, on the other, the denazification process (which "disappeared" Hitler's remains and all Nazi iconography), the reasons for believing his death had not occurred quickly becomes apparent.

In 2016, a survey revealed that 52% of people would like their Facebook page to be updated after they died. This could take the form of replies to those leaving sympathy messages on their page, or the regular reposting of photographs, videos, or other memories of the "dead" person.[26] As our social lives and our social media lives are becoming increasingly interchangeable, online presence after death is taking on palpable and interactive forms. The loss of a loved one is now, perhaps more than ever, paradoxically wrapped up in their throbbing presence, whether through digital recordings, virtual reality, or automated post-mortem activities. Is death now impossible? In Chap. 7, "Death's Impossible Date", Douglas J. Davies explores some of the philosophical intricacies of the question "when is death?" Raising the themes of animacy, grief, burial, and the "mortality paradox", Davies echoes other contributors in this book by claiming that death has an impossible date because "the 'when' of death is not coeval with 'the time of not being'".

In Chap. 8, "The Legal Definition of Death and the Right to Life", Elizabeth Wicks examines the legal implications of modern means of ascertaining death and life. Every day in hospitals and courts, medical and legal authorities are making profound and difficult decisions about the biological status and destination of vulnerable bodies. In October 2016, for instance, a terminally ill 14-year-old girl won a legal fight to have her body cryogenically preserved after death because she "'wanted to live longer'" and have a chance "'to be cured and woken up'".[27] Focusing on debates surrounding the issue of brain death in her contribution, Wicks raises the tension between our legally enforceable right to life and the state's lawful withdrawal of life-sustaining treatment in the case of people in a persistent vegetative state. This withdrawal is, she concludes, "sometimes ethically appropriate, morally good, and respectful of the human being's rights". The right to life, then, "is always limited, both in terms of state obligations and its application to mortal beings".

An execution is a usually a strictly timed event: a sentence of death is passed, the defendant's days are "numbered", and the execution itself follows a sequence of rites and behaviours. The "when" of the condemned criminal's death, then, is known for certain. In Chap. 9, "The Last Moment", Jonathan Rée focuses on this disturbing kind of death timing. He suggests that people experience a particular thrill and empathy when they imagine the final minutes and seconds of a person's life. Surveying examples from literature, philosophy, and the history of crime, Rée finds that the evolution of execution narratives has a lot to do with social attitudes about capital punishment and, in particular, an urge to think about our own last moments of life.

Finally, in Chap. 10, Thomas Laqueur's afterword focuses on the possibility of future breaths—determined by the apnea test in brain death situations—as a way to think about the end of life. Brain dead people may subsist for decades attached to ventilators and participate in the same biological milestones as everyone else (puberty, pregnancy, death). Despite the expanding chronologies of the "living dead" through science and technology, however, Laqueur argues that the "when" of death starts at the time when it is shown that a person will never breathe again without artificial assistance. This particular death sentence, of course, does not discount the reality that becoming dead also takes time in other, non-biological ways of thinking.

In its movement from history and literature, to philosophy and ethics, the contributions in this book attest to a pervasive dynamic between finality and continuance, between death as a concrete biological event and death as a social negotiation. The question we have addressed is inherently interdisciplinary. It will continue to fascinate scholarly and lay audiences alike, because death timings allow us to make sense of who we are as individuals and societies in the midst of time, shorn between long memories and imagined futures on the one hand, and a single irrevocable destiny on the other.

Notes

1. See Dorthe Refslund Christensen and Rane Willerslev eds., *Taming Time, Timing Death: Social Technologies and Ritual* (Farnham, 2013)
2. See http://www.criminalcorpses.com.
3. See Elizabeth T. Hurren, *Dissecting the Criminal Corpse: Staging Post-Execution Punishment in Early Modern England* (Basingstoke, 2016); Shane McCorristine, *William Corder and the Red Barn Murder: Journeys of the Criminal Body* (Basingstoke, 2014); Sarah Tarlow and Zoe Dyndor, "The Landscape of the Gibbet", *Landscape History*, 36:1 (2015), pp. 71–88; Owen Davies and Francesca Matteoni, "'A Virtue Beyond all Medicine': The Hanged Man's Hand, Gallows Tradition and Healing in Eighteenth- and Nineteenth-Century England", *Social History of Medicine*, 28:4 (2015), pp. 686–705; Richard Ward ed., *A Global History of Execution and the Criminal Corpse* (Basingstoke, 2015); Peter King

and Richard Ward, "Rethinking the Bloody Code in Eighteenth-Century Britain: Capital Punishment at the Centre and on the Periphery", *Past and Present*, 228 (2015), pp. 159–205.
4. Samuel Carter Hall and Anna Maria Hall, *Ireland: Its Scenery, Character, &c.*, Vol.1 (London, 1841), p. 221.
5. Ibid., ff. p. 222.
6. John O'Donovan, "Notes to Extracts from the Journal of Thomas Dineley, etc.", *Journal of the Kilkenny Archaeological Society*, 5 (1858–1859), p. 31.
7. Cited in Linda-May Ballard, "Fairies and the Supernatural on Reachrai". In Peter Narváez ed., *The Good People: New Fairylore Essays* (New York, 1991), p. 58.
8. The School's Collection, National Folklore Collection, Ireland, University College Dublin, Vol. 0102, pp. 143–144, http://www.duchas.ie/en/cbes.
9. Ibid., p. 145.
10. James Mooney, "The Funeral Customs of Ireland", *Proceedings of the American Philosophical Society*, 25:128 (1888), p. 267.
11. Seán Ó'Súilleabháin, *Irish Wake Amusements* (Dublin, 1967), pp. 171–172.
12. Cited in Patricia Lysaght, "Hospitality at Wakes and Funerals in Ireland from the Seventeenth to the Nineteenth Century: Some Evidence from the Written Record", *Folklore*, 114 (2003), pp. 406–407.
13. Patricia Lysaght, "*Caoineadh os Cionn Coirp*: The Lament for the Dead in Ireland", *Folklore*, 108 (1997), p. 69.
14. Joseph Cunnane, "The Funeral Service: I: Death and the Removal of the Remains", *The Furrow*, 8:10 (1957), pp. 631–643.
15. Maria Edgeworth, *Castle Rackrent; An Hibernian Tale, etc.*, 3rd ed. (London, 1801), p. 212.
16. Ibid., p. 213.
17. See E. Estyn Evans, *Irish Folk Ways* (London, 1957), Plate 14.
18. The School's Collection, Vol. 0122, p. 329.
19. Ó'Súilleabháin, pp. 66–67.
20. Ibid., p. 67.
21. Henry Morris, "Irish Wake Games", *Béaloideas*, 8:2 (1938), p. 121.
22. O'Donovan, p. 32.
23. "A 'Wake' in the Borough", *Birmingham Daily Post*, 19 December 1871.
24. See Gearóid Ó Crualaoich, "The 'Merry' Wake". In James S. Donnelly Jr., and Kerby A. Miller eds., *Irish Popular Culture, 1650–1850* (Dublin, 1998), pp. 173–200.
25. Thomas W. Laqueur, *The Work of the Dead: A Cultural History of Mortal Remains* (Princeton, New Jersey, 2015), p. 10.

26. "Facebook Users want to 'Check-in' from Beyond the Grave", *BBC News*, http://www.bbc.co.uk/news/uk-38154677?SThisFB.
27. "Terminally Ill Teen Won Historic Ruling to Preserve Body", BBC News, http://www.bbc.co.uk/news/health-38012267.

AUTHOR BIOGRAPHY

Shane McCorristine is an interdisciplinary historian with research interests in the history of Arctic exploration, the cultural geography of extinction, and the place of the supernatural in the modern world. Between 2013 and 2015 he was a Wellcome Postdoctoral Fellow on the "Harnessing the Power of the Criminal Corpse" project at the University of Leicester. His most recent book is *William Corder and the Red Barn Murder: Journeys of the Criminal Body* (Palgrave Macmillan, 2014).

Open Access This chapter is licensed under the terms of the Creative Commons Attribution 4.0 International License (http://creativecommons.org/licenses/by/4.0/), which permits use, sharing, adaptation, distribution and reproduction in any medium or format, as long as you give appropriate credit to the original author(s) and the source, provide a link to the Creative Commons license and indicate if changes were made.

The images or other third party material in this chapter are included in the chapter's Creative Commons license, unless indicated otherwise in a credit line to the material. If material is not included in the chapter's Creative Commons license and your intended use is not permitted by statutory regulation or exceeds the permitted use, you will need to obtain permission directly from the copyright holder.

CHAPTER 2

Being Dead in Shakespearean Tragedy

Mary Ann Lund

In his novel *Being Dead* (1999), Jim Crace intertwines the life of a married couple, both zoologists, with an account of what happens after their deaths. We learn at the beginning of the novel that they have been murdered on a beach, and their bodies lie undiscovered for some time. Their physical decomposition is described in details that are simultaneously scientific and loving. The unfurling of this story is prefaced by an epigraphic verse, "The Biologist's Valediction to his Wife" (purportedly by Sherwin Stephens, a pseudonym for Crace), which provides a quite different perspective on being dead. Where the novel is tender, the poem is comically stark. The speaker of the poem declares to his wife that there is no chance of eternal life: "You're dead. That's it. *Adieu*. Farewell". Death is an entirely one-way process of "Rot, Rot, Rot,/As you regress, from *Zoo.* to *Bot.*". The wife's being is eroded by putrefaction, and while he assures her that he will grieve for her, he also dismisses the value of so doing, since:

M.A. Lund (✉)
University of Leicester, Leicester, UK
e-mail: maejl1@leicester.ac.uk

© The Author(s) 2017
S. McCorristine (ed.), *Interdisciplinary Perspectives on Mortality and its Timings*, Palgrave Historical Studies in the Criminal Corpse and its Afterlife, DOI 10.1057/978-1-137-58328-4_2

...Grieving's never

Lengthened Life

Or coaxed a single extra Breath

Out of a Body touched by Death.[1]

No strength of feeling will move the addressee back from *"Bot."* to *"Zoo."*, even momentarily. The facts of biological death and decay are bare and brutal. Crace's verse is an anti-elegy that seems to express mourning while dispassionately describing the physical processes of death. Yet, alongside this rather shocking viewpoint, what is notable about this poem is the way the woman is addressed: she is both informed that "You're dead" and called "Departing wife". We imagine that the wife is about to die, but the biologist is imagining her as dead already as he gives her this helpful cold comfort. There is an odd paradox in this flippant bit of verse: even while he is envisaging her as no more than "manure", there is still a sense of being attached to her (and even the botanical classification gives her some life).

This chapter looks at the speech acts that denote and surround death, and more specifically, the part they play in enacting and indicating death timings on the English Renaissance stage. According to the groundbreaking language theory propounded by J.L. Austin, speech does not only *say*, but also *does*: it performs something through the act of speaking (in his terminology, it has perlocutionary force).[2] For example, when a member of the royal family says the words "I name this ship..." in the appropriate circumstances, the ship becomes named through the speaking of the words. To return to my earlier example, the phrase "You're dead" is not a merely descriptive act. It is also a performative utterance that is, moreover, heavily context-dependent. When the speaker of Crace's poem says, "You're dead", he is imagining his "departing" wife as already departed, and in a sense, is creating her as socially dead before she is biologically so. At the same time, his choice of words militates against his ruthlessly biological reading of death as simple physical decay and taxonomical change. By addressing her as dead, he allows the subject to persist, acknowledging that there is still a "you", not an "it" or even a "she". As we shall see, the tragic drama of William Shakespeare (1564–1616) has a similar fascination both with the experience of dying, and with the paradox expressed in the notion of simultaneously being, and being dead.

The Borderlands of Death: Renaissance Tragedy

The drama of late sixteenth- and early seventeenth-century England is heavily occupied not only with deaths, but with the depiction and vocalisation of the stages around them. Most obviously, the afterlife speaks and walks on the stage in the supernatural manifestations of ghosts hungry for vengeance—a staple element of the revenge tragedy tradition. In the Shakespearean canon, *Richard III*, *Julius Caesar*, *Hamlet*, and *Macbeth* all feature revenants of some kind who communicate with the living, and even play a significant part in the plot (in the case of *Julius Caesar*, the title character himself returns from the dead to haunt his own play). Such memorable, and sometimes lurid, figures are not the only dramatic reflection on mortality to command attention. Thomas Kyd's (1558–1594) highly influential revenge play *The Spanish Tragedy* (a story framed by the narrative of a ghost seeking revenge for his murder) exploits the full dramatic potential of deaths that are not what they seem. At the climax of the tragedy, a living avenger, Hieronymo, stages his own play and invites other characters to participate. What they do not realise is that the guilty parties will be killed for real, when they naturally assume that they are only acting. It only gradually becomes apparent to the audience that the play-within-a-play has become enacted in truth: the dead characters are not going to stand up again, and the blood from their wounds is not fake. Yet they will stand up, of course, and the blood is fake, because they are actors playing characters who play characters: the fake-real-fake-real feint Kyd uses is as much a self-referential game as is the custom of boy actors playing female characters, who in turn dress up as men.

Renaissance dramatists experimented with, and challenged, the audiences' expectations and assumptions about how death occurs on stage. If we witness a character being strangled and see her immobile body lying on the stage or concealed behind a curtain, we assume that she is dead and gone, even though we know, and perhaps can even see, that the actor playing her is still breathing. As we shall later see, Shakespeare uses various conventions to indicate death, in particular verbal cues. Yet conventions can also be overturned, to complicate our notions of death timings. In some cases, the audience is privy to the secret that a character is not dead despite seeming to be so, a knowledge kept from other characters on stage. For example, in *Romeo and Juliet* we witness Friar Laurence explaining to Juliet his plan to reunite her with her exiled lover Romeo: she should drink a potion to suppress her physical responses, so that "No warmth, no breath shall testify thou livest"; each part of her body "Shall, stiff and stark and cold, appear like death" (*Romeo and Juliet*, 3.5.98, 103).³ When the unknowing Romeo breaks open her tomb, we know that he has misread the signs given by her body, and when he poisons himself—"Thus with a kiss I die!" (5.3.120)—we are acutely aware of the horrendous nature of this tragedy; Juliet will soon rise up from her drugged sleep-as-death to find the still-warm corpse of her new husband Romeo, and she will herself commit suicide.

Juliet's second death by stabbing is genuine and observed as such (the watchman pronounces her "bleeding, warm, and newly dead,/ Who here hath lain this two days" (5.3. 174–175)), but not all women in Shakespeare's tragedies pass away so swiftly or evidently. A strange contrast to Juliet's case is that of Desdemona, smothered by her husband Othello in a jealous rage. In such a death, there are no immediately readable signs like Juliet's fresh blood, especially since, as Othello himself has observed earlier, Desdemona has "whiter skin [...] than snow,/And smooth as monumental alabaster" (*Othello*, 5.2.4–5). Even when she is alive and in good health, Othello imagines her as a dead figure like that on a stone tomb, perhaps to lessen his guilt for what he is about to do. Her physical self presents a dramatic contrast to Juliet, in whom, even when she is drugged and seemingly "dead" within a vault, Romeo notices that "Beauty's ensign yet/Is crimson in thy lips and in thy cheeks" (*Romeo and Juliet*, 5.3. 94–95). When Othello first smothers Desdemona, he does not fully succeed, and remarks as much—"Not dead? Not yet quite dead?" (*Othello*, 5.2.95)—continuing until he can pronounce firmly that "She's dead" (5.2.100). But shortly afterwards, we discover that we cannot trust his reading of the external signs of mortality, when we hear her voice from behind the bed-curtains: "O, falsely, falsely murdered!" (5.2.126). Although we may initially conclude that she is a ghost, we soon realize that she has indeed revived, just for long enough to proclaim her innocence, and also to exonerate, rather than accuse, her husband. Her maid, Emilia, asks her "who hath done this deed?"; Desdemona's final words are "Nobody, I myself" (5.2.132, 134).

Othello is not the only example in Renaissance drama where a suffocated woman temporarily revives. Shakespeare's Jacobean contemporary John Webster (1580–1634) takes the idea even further: when the eponymous heroine of his play *The Duchess of Malfi* is strangled on the orders of her brother, she revives over a hundred lines after she has seemingly "died". The man who has carried out the murder, Bosola, realises that "She's warm, she breathes", "her eye opes", and he tells her that her husband is still alive, allowing her to die happily: "Mercy" is her final word.[4] This revivification of women may, as Kaara Peterson convincingly argues, be connected to the early modern medical understanding of the female disorder *hysterica passio*, in which women "display the symptoms that mimic death", a representation reflecting a wider cultural anxiety that "women's bodies cannot be trusted to reflect their most fundamental status as living beings".[5] Perhaps Renaissance audiences were more sceptical about whether a female character's death was definitive unless clear signs were exhibited. Moreover, such a medical understanding allows, and even enables, the fantasy that grief can coax a single extra breath out of a body touched by death; a particular preoccupation of Shakespeare's late drama. In his play *Pericles*, a coffin washed up on the shore is opened to reveal a still-pink body of Thaisa, buried at sea, who is revived back into life by a skilful doctor, and at last reunited with her husband; in *The Winter's Tale*, there is another pink female form—the statue of the wronged late Queen Hermione—that miraculously steps off its plinth and reveals itself to be a breathing woman. The borderlands of death allow redemption as well as consolation.

BECOMING DEAD IN SHAKESPEARE

Renaissance drama reminds us that the boundaries between life and death can be paper-thin, that observed signs may not be sufficient, and that characters and actors inhabit worlds of being and not-being. The most striking reminder of death as merely performance is the jig at the end of a play, a traditional coda to the action. There may be a pile of bodies, or a funeral procession, but in a Renaissance tragedy that is not "it. *Adieu*. Farewell". As the play's action ends, the actors re-enter to dance, an experience recounted by the Swiss traveller Thomas Platter, who visited London in 1599: "On the 21st of September, after dinner, at about two o'clock, I went with my party across the water; in the straw-thatched house we saw the tragedy of the first Emperor Julius Caesar, very pleasingly performed, with approximately fifteen characters; at the end of the play they danced together admirably and exceedingly gracefully, according to their custom, two in each group dressed in men's and two in women's apparel".[6] There is something of the *danse macabre* to this idea of reanimated corpses, but the revival of this tradition in the

Globe Theatre in recent years demonstrates how dancing provides a sense of release for actors and audience, an emotional and aesthetic complement to the catharsis of watching tragedy.

The question "when is death?" finds a particularly nuanced response in two of Shakespeare's plays: *Hamlet* and *King Lear*. Both tragedies are invested not only in death, but in the process of becoming dead, and do so through their male protagonists. I have suggested that the phrase "You're dead" seems paradoxical, and we hear an equivalent one on the Renaissance stage:

> Hamlet, thou art slain.
>
> No med'cine in the world can do thee good.
>
> In thee there is not half an hour of life;
>
> (*Hamlet*, 5.2.266–268)

The words are spoken by Laertes, who has killed Hamlet in revenge for the murder of his father, Polonius, and the death of his sister, Ophelia. Having challenged Hamlet to a duel, Laertes poisons the tip of his foil so that even a flesh wound will kill his rival; during a scuffle in the swordplay, the foils are accidentally switched, and both men are wounded and fatally poisoned. Laertes is able to say, "thou art slain" not only because he has administered the poison, but also because he too is on the verge of death: "Lo, here I lie,/Never to rise again" (5.2.271–272), he says. His own internal feeling of approaching death becomes projected onto Hamlet, while Hamlet can see his own death in Laertes. Laertes becomes a living emblem of the traditional tomb inscription, "*Eris quod sum*" ("what I am, you will be").

Hamlet's recognition of what Laertes has told him reveals itself in a phrase even more paradoxical than "thou art slain": "I am dead" (5.2.285, 290). Hamlet says the phrase twice within a single speech to his friend Horatio, just after they have witnessed Laertes dying. At first, we might think that Hamlet is indulging in a touch of melodrama by repeatedly drawing attention to his own process of dying. Certainly, his death is among the most protracted in the Shakespearean canon: he dies 60 lines after he has been wounded, which is twice as long as it takes Laertes to die from the same poison. As is often the case on the Renaissance stage, theatrical demands and medical theory meld in the behaviour of characters' bodies. Hamlet dies more slowly than Laertes because he is the protagonist, and his death should take centre stage, but also because of his physiology. According to the prevailing Galenic medical understanding of the body in this period, those people with a hot or dry (sanguine or choleric) temperament were believed to be more susceptible to the workings of poison, because the natural heat in their bodies made the poison work faster, and because they were believed to have wider arteries.[7] As the preacher John Donne would later put it, "Poyson works apace upon cholerike complexions".[8] The hot-headed Laertes dies before the melancholic Hamlet, whose cold humoral temperament preserves him longer, just as the dramatist allows him the prolonged time to give his "dying voice" (5.2.340), being acutely aware of his own theatricality in front of the "audience to this act" (5.2.320).

What does Hamlet mean when he says "I am dead"? Perhaps it is just another way to say, "I am dying"; "I'm as good as dead"; or "I'm done for". But this phrase is hardly colloquial. In a search across English writing of the same period, instances when someone says, "I am dead" (usually in drama) are typically preceded by "when", or "imagine", words that make clear that the speaker is predicting or imagining, not inhabiting death as Hamlet does.[9] Hamlet's phrase appears to be unique on the English stage. I would argue that the mirroring of Laertes' phrase is more than simply a substitute for saying that he is dying. Hamlet, the student, is well versed in ontology, and has already revealed his reflections on what it is like to be dead, "not to be":

> ...To die, to sleep.
>
> To sleep, perchance to dream. Ay, there's the rub,
>
> For in that sleep of death what dreams may come
>
> When we have shuffled off this mortal coil
>
> Must give us pause. (3.1.66–70)

Hamlet has spent most of the play's action thinking about and preparing himself, firstly, to revenge the murder of his father by his uncle, King Claudius, and secondly to die. Within 10 lines he has finally achieved both: he has at long last successfully taken on the role of avenger, killing Claudius (who, incidentally, takes a mere five lines to die) with a combination of poisoned implements, and he has also become the revenged: he has taken on the status of being dead, while he is still conscious.

There are further dimensions to Hamlet's extraordinary speech act "I am dead". The editors of the Arden edition of *Hamlet* note of these lines that "If Hamlet is already 'dead' when he kills the King, this may be Shakespeare's solution to the moral dilemma of the blood-guilt of the successful revenger".[10] I think this explanation fails to account for Hamlet's own deep investment in the theology and ethics of revenge; for Hamlet to absolve himself of guilt through this speech act seems a solution too neat and untroubled for him. Hamlet's expression of simultaneously being, and being dead, is perhaps better understood as an articulation of death being experienced in different timings, rather than as a single event. From a legal and medical (if not an ethical) perspective, he may see himself already as a dead man. When Hamlet hurts the king with his poisoned foil, there is a general cry of "Treason, treason!" (5.2.275). Hamlet has not been tried for treason, but he has fully taken on the responsibility for killing a king, and there is hence a legal sense in which Hamlet is already dead: those who are condemned for treason (not necessarily through trial) lose all their civil rights and capacities, and forfeit their estates and their ability to transmit property to descendents.[11] Hamlet himself characterizes death as "this fell sergeant" who is "strict in his arrest" (5.2.288–289), exploiting the idea of himself as a prisoner. Furthermore, a medical symptom may overlay his legal and philosophical awareness. He tells Horatio that "the potent poison quite o'ercrows my spirit" (5.2.305), an expression of the moment of death as it is being experienced. The words "I am dead" may be a manifestation of that, of the symptom *angor animi*: that is, "the sense of being in the act of dying, differing from the fear of death or the desire of death" encountered, for example, in angina patients.[12] "Angor animi" might be translated as "anguish of the spirit", and in Shakespeare's time, the term referred to mental anguish and melancholy, of the kind to which Hamlet was well-accustomed. Here, I suspect, he is also experiencing that more modern sense of the term: feeling death rather than just contemplating it.

Textual Deaths in *Hamlet*

If we re-pose the question "when is death?" in *Hamlet*, we should look further at the protagonist's final words. There is no ambiguity about his passing moment in the way that there is about Desdemona, or the

Duchess of Malfi—no trick where an extra breath is coaxed out of him after he seems to have breathed his last. There is, however, an interesting textual variation that serves to highlight the ways in which death can be signalled, both on stage and in a reading of the printed text. Hamlet experiences death in three different ways, according to which version of the play is consulted. The first time that *Hamlet* appeared in print, in 1603 (the First or so-called "bad" Quarto), was in a version significantly shorter and markedly different from later editions; possibly it was a reconstruction of Shakespeare's play from memory by an actor or audience member, a "bootleg" edition with some notably garbled passages.[13] In this version, Hamlet's words are (to us) unfamiliar, but conventionally final:

> What tongue should tell the story of our deaths,
>
> If not from thee? O my heart sinckes *Horatio*,
>
> Mine eyes have lost their sight, my tongue his use:
>
> Farewel *Horatio*, heaven receive my soule.   *Ham. dies.*[14]

Hamlet asks his friend Horatio not to commit suicide and accompany him in death, but instead to remain alive and bear testimony to what has happened. As the poison takes effect, the First Quarto Hamlet acts as a kind of witness to the physical symptoms of dying he is experiencing. The sinking heart and loss of senses are conventional medical indicators that death is near: one manual lists "difficultie of breathynge, dimnesse of sight, dulness of sense" among the warning signs.[15] His articulation of physical experiences and symptoms as a representation of an unobservable, interior state is also a dramatic convention, being characteristic of the Senecan mode to which Elizabethan tragic theatre was indebted.[16] But it is a firmly Christianized Senecanism, as Hamlet's final words indicate in the form of a spiritual piety, "heaven receive my soul". The stage direction confirms the obvious: "*Ham. dies*". The First Quarto, then, gives a triple set of indications of the moment of death: in verbalized medical symptoms, in spiritual preparation, and in an instruction to the actors (and, by extension, to the reader of the printed text).

Yet, this version is mostly rejected by editors; in the Second Quarto of 1604, he dies quite differently, or indeed, not at all (textually, at least). He starts to give a message for the Norwegian prince Fortinbras (whom Hamlet would like to take over the rule of Denmark), but he stops halfway through his sentence: "So tell him, with th'occurrants more and lesse/Which have solicited, the rest is silence".[17] The abrupt last phrase is unaccompanied by any stage direction, and it is his friend Horatio's words in response that make clear that he has died: "Now cracks a noble hart, good night sweete Prince,/And flights of Angels sing thee to thy rest". In the First Folio, the collection of plays published 7 years after Shakespeare's death, Hamlet's last lines are nearly identical, but the text gives him a death groan "O, o, o, o" and a stage direction: "*Dyes*".[18]

Although the versions are different, Hamlet's ending is unambiguous. There is no indication that he lingers on after his final line. Indeed, between the three editions we can observe a combination of methods by which death is indicated at a textual level in an English Renaissance play. The clearest is the stage direction: "*Dyes*". This, like all stage directions, is used sporadically in printed texts; it is altogether common to find no such stage direction, which means that the modern editor of a play usually puts one in (and sometimes has to make an educated guess about where it occurs). A death groan in a printed text ("O, o, o, o") must be a shorthand for the combination of unscripted sound and gesture with which tragic actors such as Richard Burbage performed their death throes. It is worth remembering that the sightlines of public playhouses, with a raised stage and standing spectators in the yard, make different demands of a performed death from those in modern theatres; although the traveller Thomas Platter praised the layout of English playhouses where "everyone can well see everything",[19] an actor lying down on the stage would need to die vocally as well as visually, in order to communicate to those standing round the stage. A resounding final line is another clear way of doing this. A rhyming couplet is a traditional way to end a long speech in blank verse, and is also commonly used to end a character's life. Brutus in *Julius Caesar* meets a self-determined death by stabbing—the "Roman way" of suicide—with a similarly self-determined couplet, and Othello does exactly the same (stabbing is generally a quick death in Shakespeare, and characters tend to die within two lines, with the exception of Antony in *Antony and Cleopatra*, who bungles his suicide and takes a horribly protracted hundred lines to die, split over two scenes). And the last indicator of death (although the least reliable) is a pronouncement by another character. This is often, though, where ambiguity can creep in.

Produce the Bodies: *King Lear*

I will finally turn to *King Lear*, another of Shakespeare's tragedies where the promised end is textually, medically, legally hard to pinpoint. When the aged and mentally frail Lear enters the stage for the last time, it is to the shock of anyone who knows only the historical source of the play. In the English chronicle histories which Shakespeare drew on as a source, the ancient English king Leir is outlived by his youngest daughter Cordeil. But in Shakespeare's version, Lear comes in carrying her seemingly lifeless body: "Howl, howl, howl [...] she's gone for ever" (*King Lear*, 5.3.255, 257). He declares authoritatively that "I know when one is dead, and when one lives,/She'd dead as earth", yet even so, he immediately demands the materials for testing life, a mirror and a feather, and claims that "This feather stirs, she lives" (5.3.263). In most productions, and indeed in many editions of the play, the possibility that Cordelia is still alive is rejected, and Lear's behaviour is taken as a poignant exhibition of grief and delusion.[20] Yet the possibility is still hinted at, and once again the play-text supports this ambiguity. Even more than in *Hamlet*, there are significant variations between versions—in this case the First Quarto (1608) and the First Folio (1623)—but both have the same stage direction: "*Enter Lear with Cordelia in his armes*", not "with the body of Cordelia".[21] This phrasing is particularly telling since, a little earlier, Lear's other daughters, one poisoned, the other stabbed, are carried on, and the stage direction specifies that "*The bodies of Gonerill and Regan are brought in*". The stage directions thus appear to distinguish between the representations of two who are obviously corpses, and one who may or may not be. Lear's pain is all the more prolonged if signs of Cordelia's life appear and recede, a reminder that biological death is by no means instantaneous. Shakespeare fills the stage with bodies—unusually, even those who have died offstage are brought in—so that, by the end, the royal father and the three daughters for whom he divided his kingdom are all seen lying together and carried out in the final ritual of tragedy; the First Folio even specifies the musical accompaniment: "*Exeunt with a dead march*".[22]

I have suggested that Cordelia's death timing is ambiguous where that of her sisters Goneril and Regan is not—they are identified as "bodies"—but the latter case is worthy of further scrutiny. The last scene of *King Lear*, like the character Hamlet, shows a preoccupation with the liminal state of being both alive and dead, one which all of the Lear family occupy at some point. Early on in the scene, Goneril's husband Albany takes charge of the court, as the full wickedness of a plot involving the two sisters Goneril, Regan, and their lover Edmund is revealed. Albany directly charges Edmund, as well as his own wife Goneril: "I arrest thee/On capital treason, and in thine attaint/This gilded serpent" (5.3.83–85). Attaint or attainder signifies "the legal consequences of being condemned for treason, i.e., death and forfeiture of wealth and honour, as if the blood were [...] irretrievably stained".[23] Edmund, having been attainted for "heinous, manifest and many treasons" (5.3.93), is ordered to a trial by combat, which he loses, but Goneril quibbles with the legality of this. When Albany confronts her with the paper evidence of her infidelity, she responds that "the laws are mine, not thine./ Who can arraign me for't?" (5.3.156–157). Albany has married into her power and lands, and she is moreover a queen and, in her eyes, above the law.

The conflict between Albany and Goneril thus becomes a struggle over different assertions of legal authority, and, by extension, over the control of life and death. By pronouncing an attainder on Edmund and Goneril, Albany is staking a claim for his power to pronounce them both legally dead, before they are biologically extinct. When Goneril later poisons her sister and stabs herself, Albany orders attendants to "Produce the bodies, be they alive or dead" (5.3.229). Albany's careful wording reasserts the authority that Goneril has disputed: the sisters take on the status of "bodies" through *habeas corpus*, and also as people attainted of high treason. The equivalent is "bring up the bodies", as used in the novel of the same name by Hilary Mantel (2012), a reference to an order to bring Anne Boleyn's supposed lovers to court; the same language was used in the trials for high treason of Henry, Duke of Suffolk in 1554 and Patrick O'Collun in 1594.[24] Goneril and Regan are in all senses dead when they are brought in, and problems of jurisdiction have been sidestepped; for Albany, it is "This judgement of the heavens that makes us tremble" (5.3.230).

Ideas of medical and legal forms of death are thus raised through the demise of Lear's three daughters, but at last it is Lear's own death that shows us the full tragic force of the "when is death?" question. As Emily R. Wilson puts it, "[t]he exact moment of his passing is obscure:

the other characters on stage seem puzzled about when Lear dies and wary of saying prematurely that 'He is gone indeed'. Lear has already experienced practical death so many times that the moment of his physical death seems unimportant". Lear is, in Wilson's words, an example of "tragic overliving".[25] In the First Quarto, he gives a dying groan ("O, o, o, o") and gives an *envoi* to a world of pain in his last words, "Breake hart, I prethe breake"; there is no stage direction to mark the precise timing of his death.[26] In the later First Folio, there is a clear stage direction "*He di[e]s*",[27] but this revised text also erases the other formal signs of dying in Lear's own speech, and the line "Breake heart, I prythee breake" becomes an expression not of a dying moment, but of pain and grief spoken by another character who witnesses Lear's death. Lear, indeed, is looking at his dead daughter's lips, "Looke there, looke there", when the stage direction indicates that he dies—fading away, perhaps in hope that there is a sign of life.

Shakespeare's tragic drama is deeply invested in the question of how death timings occur, and in the issues of power that surround them: medically, legally, philosophically, and emotionally. Because they are performed, moments of passing become a participatory process in which the character (or actor-as-character), other characters on stage, and the audience play a role in enacting them. The words spoken by the witnesses to King Lear's death, "O let him pass" and "he is gone indeed" (5.3.312, 314), do not only recognise the stages of death, but also enjoin audience members or readers to let Lear pass—perhaps in a way akin to Prospero's epilogue to *The Tempest*, when he asks the audience to "release me from my bands/With the help of your good hands" (Epilogue, 9–10). For the process of becoming dead in a tragedy extends beyond the final lines, into the solemn music of the funeral march, the procession that carries the bodies off stage, and finally the actors' re-entry, revived, for the jig and the audience's applause. The medical and legal dimensions of becoming dead, Hamlet's testimony of death as it is being experienced, and Lear's fading moments all connect us to that communal experience of seeing and feeling staged death—part of the Aristotelian understanding of cathartic tragic drama.[28] The statement to which tragedy always gestures, in the paradoxically pleasurable way that drama does, is not "you're dead" or "I am dead", but "we are dead".

Notes

1. Jim Crace, *Being Dead* (London, 1999), n.p.
2. See J.L. Austin, *How to Do Things with Words: The William James Lectures Delivered at Harvard University in 1955* (Oxford, 1962).
3. References to Shakespeare's plays are included in-text. *Romeo and Juliet*, *Othello*, *Hamlet*, and *The Tempest* are referenced from William Shakespeare, *The Complete Works*, Stanley Wells et al. eds. (Oxford, 1988); *King Lear* is referenced from William Shakespeare, *King Lear*, R. A. Foakes ed. (London, 2003).
4. *The Duchess of Malfi*, 4.2.341, 345, 351 in John Webster, *Three Plays* D.C. Gunby ed., Rev ed. (Harmondsworth, 1995).
5. Kaara L. Peterson, *Popular Medicine, Hysterical Disease, and Social Controversy in Shakespeare's England* (Farnham, 2010), p. 72.
6. Ernest Schanzer, "Thomas Platter's Observations on the Elizabethan Stage", *Notes and Queries*, 201 (1956), p. 466.
7. Ambroise Paré, *The Workes of [...] Ambrose Parey* (London, 1634), sig. 3T5^r.
8. John Donne, *The Oxford Edition of the Sermons of John Donne, Vol. 12: Sermons Preached at St Paul's Cathedral, 1626*, Mary Ann Lund ed (Oxford, 2017), p. 28.
9. Search conducted via Early English Books Online, accessed 1 March 2016, http://eebo.chadwyck.com.
10. William Shakespeare, *Hamlet*, Ann Thompson and Neil Taylor eds. (London, 2006), p. 457.
11. See John Bellamy, *The Tudor Law of Treason* (London, 1979).
12. Henry Marsh, *Do No Harm: Stories of Life, Death and Brain Surgery* (London, 2014), p. 74.
13. See Shakespeare, *Hamlet*, pp. 74–86.
14. William Shakespeare, *The Tragicall Historie of Hamlet Prince of Denmarke* (London, 1603), sig. I3^v.
15. Christopher Langton, *A Very Brefe Treatise, Ordrely Declaring the Pri[n]cipal Partes of Phisick* (London, 1547), sig. k5^v.
16. See further Gordon Braden, *Renaissance Tragedy and the Senecan Tradition: Anger's Privilege* (New Haven, 1985); Robert S. Miola, *Shakespeare and Classical Tragedy: The Influence of Seneca* (Oxford, 1992).
17. William Shakespeare, *The Tragicall Historie of Hamlet, Prince of Denmarke* (London, 1604), sig. G1^v.
18. William Shakespeare, *Mr. William Shakespeares Comedies, Histories, & Tragedies Published According to the True Originall Copies* (London, 1623), sig. 2q1^r.

19. Schanzer, "Thomas Platter's Observations", p. 466.
20. Shakespeare, *King Lear*, p. 385, n. 254.1.
21. William Shakespeare, *M. William Shak-speare: His True Chronicle Historie of the Life and Death of King Lear* (London, 1608), sig. L3ᵛ; Shakespeare, *Mr. William Shakespeares Comedies*, sig. 2s2ᵛ.
22. Shakespeare, *Mr. William Shakespeares Comedies*, sig. 2s3ʳ.
23. Shakespeare, *King Lear*, p. 371, n. 84.
24. House of Commons, *Fourth Report of the Deputy Keeper of the Public Records* (London, 1843), pp. 245, 284.
25. Emily R. Wilson, *Mocked with Death: Tragic Overliving from Sophocles to Milton* (Baltimore, 2004), p. 126.
26. Shakespeare, *M. William Shakespeare*, sig. L4ʳ.
27. Shakespeare, *Mr. William Shakespeares Comedies*, sig. 2s3ʳ.
28. See Aristotle, *The Poetics of Aristotle: Translation and Commentary*, Stephen Halliwell trans. (London, 1987).

Author Biography

Mary Ann Lund is Lecturer in Renaissance English Literature at the University of Leicester. She is the author of *Melancholy, Medicine and Religion in Early Modern England: Reading 'The Anatomy of Melancholy'* (Cambridge University Press, 2010), and editor of volumes 12 and 13 of *The Oxford Edition of the Sermons of John Donne*.

Open Access This chapter is licensed under the terms of the Creative Commons Attribution 4.0 International License (http://creativecommons.org/licenses/by/4.0/), which permits use, sharing, adaptation, distribution and reproduction in any medium or format, as long as you give appropriate credit to the original author(s) and the source, provide a link to the Creative Commons license and indicate if changes were made.

The images or other third party material in this chapter are included in the chapter's Creative Commons license, unless indicated otherwise in a credit line to the material. If material is not included in the chapter's Creative Commons license and your intended use is not permitted by statutory regulation or exceeds the permitted use, you will need to obtain permission directly from the copyright holder.

CHAPTER 3

"A Candidate for Immortality": Martyrdom, Memory, and the Marquis of Montrose

Rachel Bennett

A salient theme in this book is that death neither has an entirely static definition nor does its timing always have a discernible chronology. To quote Thomas Laqueur, natural death is something that happens in an instant but "becoming really dead…takes time".[1] This chapter engages with the central question of "When is Death?" by looking at the death of one man in particular, James Graham, the 1st Marquis of Montrose (1612–1650). Montrose played a prominent role in the early part of the mid-seventeenth century religious and military conflicts between the Covenanters in Scotland and the Stuart monarch of the three kingdoms of England, Scotland, and Ireland. His place in the history of the struggle was complex because he initially supported the Covenanting cause before switching to support King Charles I. This chapter demonstrates that Montrose's death had multiple timings and that the use of his body highlights an

R. Bennett (✉)
University of Warwick, Coventry, UK
e-mail: rachelelizabethbennett@gmail.com

important interplay between issues of power and punishment, and martyrdom and memory.

First, this chapter provides a brief timeline of the events that made Montrose a key figure in the struggle between Covenant and King, and which made his death noteworthy. Second, it investigates the multiple stages of what will be defined as his "legal death". Montrose led Royalist forces in battle against the Covenanters, and for this he was outlawed, excommunicated, and attained for the crime of treason in 1644. Thus, the process of his legal death began years before he was finally captured and publically executed in Edinburgh in 1650. This section also shows how Montrose's capture began a social death, as he was paraded from northern Scotland to Edinburgh with his crimes advertised and his name degraded. Within the black catalogue of offences that carried a capital punishment in this period, the crime of treason was set apart in how it was punished. The punishment for treason extended beyond the extinction of life as the corpses of traitors were used to send out stark and richly symbolic messages. In Montrose's case, his head was spiked on top of Edinburgh's Old Tolbooth and his limbs were displayed in four of Scotland's main towns. However, as the third section of this chapter demonstrates, he was still not yet truly "dead".

After the wars of the three kingdoms, and the Interregnum of republican government, the three kingdoms were restored under Charles II in 1660. This led to a wave of Royalist sentiment, and the martyred Montrose was used to propagate the themes of loyalty and sacrifice. In 1661, the first parliament of Charles II resolved to provide some "honourable reparation" for the barbarity committed upon him.[2] Montrose's dismembered body, once used to mark out his criminality, was gathered together and given a full public funeral at the King's expense in order to mark a legal, spiritual and social rehabilitation. This, again, provokes questions over the timing of his death. The third section of the chapter charts the journeys of particular parts of Montrose's body, namely his heart and one of his arms. These body parts were not buried with the rest of his body in 1661 and they have legacies of their own. In 1925, a newspaper article discussing the potential sale of Montrose's arm argued that the desire to possess it came from "our interest in the past and a craving for the most convincing form of testimony."[3] These body parts were transformed into relics and were passed down through the generations, attracting beliefs about Montrose and also about the power of the dead body.

MONTROSE: COVENANTER AND CAVALIER

In 1638, Scottish nobles and common people alike signed the National Covenant, the purpose of which was to provide a written document stating their commitment to the Reformed religion and the principle of a church that was not controlled by the crown, but whose followers remained loyal to their king. It was signed in the Kirkyard of Greyfriar's in Edinburgh by the great Scottish lords, including the Marquis of Montrose, a young and energetic military campaigner. King Charles I alienated his Scottish subjects by reforming the liturgy and discipline of the church, leading to fears of an eventual return to popery. To the King, the Church of Scotland was greatly inferior to the Anglican Church as it lacked proper liturgy and its bishops did not have a suitably exalted status.[4] After the signing of the Treaty of Berwick in 1639, which ended the early hostilities known as the First Bishop's War, Montrose was sent to discuss the religion question with Charles I, and it was not long after this that he began to switch his allegiances. John Buchan argues that Montrose became aware that the governance of Scotland was increasingly in the hands of certain individuals, notably his great enemy the Marquis of Argyll, who he feared were committing the very breaches of the law for which they had previously condemned the King.[5]

In 1644, Montrose pledged allegiance to the King and, soon after, was appointed the Viceroy and Captain-General of Scotland. In the same year, he was attained and outlawed for treason as well as being excommunicated by the Covenant Committee of Estates, thus marking the beginning of his legal, but also his spiritual death, in the eyes of his enemies. Montrose's cavalier forces went on to achieve victories in various parts of northern Scotland, including a particularly bloody campaign in Aberdeen that blackened his reputation. In an account of the sufferings inflicted upon the Church of Scotland, the early eighteenth-century ecclesiastical historian Reverend Robert Woodrow called Montrose and others who supported the King "malignants and anti-Covenanters".[6] Even the anti-Covenanter Sir George Mackenzie referred to Montrose as a "vain-glorious butcher" for his actions in the Highlands.[7]

In contrast, Montrose fared better in later interpretations of the period. Robert Chambers, in his *History of the Rebellions in Scotland* (1828), argued that the conduct of the Covenant meant that Montrose—his hero—had to join the King to protect the rights of society from church oligarchy.[8] Other accounts praised his "heroic moderation" and the lack of malice in his military campaigns.[9] Although Montrose missed out on a heroic status in life, he enjoyed a vibrant afterlife in works of fiction that portrayed him as a romantic hero. Catriona MacDonald shows how, in the post-1745 Jacobite Rebellion period, Montrose's legacy was refashioned, along with that of the Highland clans, to exemplify Scottish national virtues in works such as Sir Walter Scott's *A Legend of Montrose* (1819).[10]

Despite winning some significant victories, Montrose encountered a lack of support in the Scottish Lowlands and he fled to Norway in 1646. It was not until 1649, following the execution of Charles I, that Montrose was restored to the lieutenancy of Scotland by the King's son and heir Charles II. In the following year, Montrose landed in the Orkney Islands with a small force of Royalists, but he failed to gain significant clan support and was defeated by Covenant forces at the Battle of Carbisdale in April 1650. He spent a few days on the run in the Scottish Highlands before coming upon a previous ally named Neil McLeod. However, instead of offering him assistance, McLeod apprehended Montrose and handed him over to the Committee of Estates for a bounty. This perceived treachery fed into the Royalist cult of martyrdom in the wake of Montrose's death. When he was apprehended by McLeod, Montrose apparently requested that he be killed quickly where he stood, rather than be handed over to his enemies.[11] However, McLeod refused this request and the Covenant made plans to use his death to make a political statement about their strength and to avenge Montrose's betrayal. The Committee of Estates wanted to bring Montrose to Edinburgh to be put to death before King Charles II arrived in Scotland and interceded to prevent the execution. Despite this urgency, they made the execution a three-day long public spectacle replete with all possible ignominy.

THE EXECUTION OF MONTROSE

Montrose had been attained, convicted, and excommunicated by the Committee of Estates in 1644 and this still stood at the time of his capture in 1650. Therefore, in this sense, his legal death had already begun years before he was physically present to hear his death sentence. Following his capture, Montrose was brought from northern Scotland to Edinburgh. While this journey was intended to bring shame to his name, and mark the continuation of his social death to Covenanters, Montrose actually received a multitude of reactions during the procession. In some places he was given food, comfortable shelter, and fine clothes thought befitting to his status. However, at places where his forces had been victorious, a herald was placed above him that proclaimed, "here comes James Graham, a traitor to his country".[12]

On 19 May 1650, Montrose was met at the city gates and conveyed with all possible ignominy to the Old Tolbooth. He was placed in a cart bare-headed and tied to a specially made seat to ensure that he was in full view of the crowd.[13] Although the distance between the gates and the Tolbooth was little more than half a mile, the procession took three hours as special stops were made along the way, including a lengthy pause outside the house from which the Marquis of Argyll and other Covenanting authorities viewed the spectacle. The city's ministers urged people to throw things at Montrose and abuse him during the procession to add further shame. However, some in the crowd were moved by his dignity and courage in the face of his ordeal, and various commentators spoke of a "tense air of sympathy and startled admiration" for him.[14]

The day after his arrival in Edinburgh, Montrose was taken before the Committee who repeated their charges of rebellion against the state and desertion of the National Covenant. He was not given a formal trial because he was already attained, and thus convicted, for his crimes. Montrose was sentenced to be hanged on 21 May at the Cross in Edinburgh on gallows that were 30 feet high. Throughout Europe during this period many noble traitors were executed by beheading, perceived as a more honourable end than hanging. However, this concession was not extended to Montrose because of the desire of the Covenanters to add even further infamy to his death. His private chaplain George Wishart wrote a biography of Montrose that favourably detailed his previous military campaigns, and this book was ordered to be placed around his neck as a reminder of his crimes. After the body was hung for three hours, it was ordered that it be cut down, beheaded and quartered. Montrose's head was to be fixed on top of the Tolbooth and his legs and arms distributed to Stirling, Glasgow, Perth, and Aberdeen. Various ministers visited him before the execution because it was stipulated that if he repented his crimes, the sentence of excommunication would be lifted. However, Montrose stated that, although he continued to hold to the Covenant he had taken, he could not support any actions against the authority of the King, to whom he pledged a greater allegiance and in whose authority he had acted. Furthermore,

Montrose apparently stated that he thought it an honour to have his loyalty remembered in Scotland's five most eminent towns.[15] During the Restoration regime, such reports of Montrose's gallantry when faced with unjust death were used to further establish his position as one of the most celebrated Royalist martyrs of the period.

On the morning of his execution, Montrose ascended the scaffold wearing fine scarlet with white gloves and silk stockings that had been provided by friends. Traditionally, criminals about to suffer the last punishment of the law were given the opportunity to address the watching crowd in order to express public penitence for their offences and to reconcile themselves with their fate. However, fears that Montrose might be rescued meant the authorities limited his access to the public. He was only permitted to address those immediately around him, one of whom recorded what he said. In his investigation of the behaviour and last dying speeches of the Jacobite rebels, Daniel Szechi argues that, during his execution, Charles I had set a precedent in refusing to publicly accept the justice of his sentence.[16] Instead of showing penitence during his last moments, Montrose reaffirmed his loyalty to God and the King, and expressed satisfaction that he was to follow in the footsteps of the martyred Charles I.[17] Reporting upon Montrose's gallant deportment, a contemporary pamphlet commented that "it is absolutely believed that he hath gained a better repute by his death than ever he did in life".[18]

After hanging for three hours, Montrose's body was cut down and his head and limbs were cut off with an axe—a scene that was met with sounds of regret from the crowd.[19] The Covenant then began to display their authority and justice by distributing the body parts. Montrose's head was spiked on top of the Old Tolbooth to mark out his treasonous criminality and to prolong his public humiliation beyond execution. One of his arms was put up at Justice Port in Aberdeen, and another was sent to Dundee. The legs were sent to Stirling and Glasgow.[20] Often, the torsos of victims thus dismembered were given to relatives for burial; this was not the case with Montrose, however. Because he had been excommunicated, his torso was buried in unconsecrated ground under the gallows on the Borough Muir. At the time, this final insult by the Scottish Kirk was considered a greater torture than the punishments inflicted upon his body in life.[21]

These post-mortem punishments were broadly consistent with other treatments meted out on the corpses of traitors during the period. The legal death sentence was designed to deny the condemned a decent burial and also to harness the power of the criminal corpse to make a political statement: this is what happens to traitors. In many cases, the dismembered body parts of the executed remained on display for years until they rotted away to nothing, or were eventually taken down without ceremony and lost to historical record. Montrose's body, however, became a vehicle to promote an entirely different political message in 1661.

"An Honourable Reparation"

In his biography of Montrose, Wishart called him a "candidate for immortality", and provided one of the earliest examples in which Montrose's death was held up as iconic in the Royalist cause. When lamenting Montrose's treatment by the Covenant, Wishart stated that his death "was not bewailed as a private loss but rather as a public calamity".[22] However, following the Restoration, Charles II's first parliament resolved to bestow upon Montrose "an honourable reparation" for the barbarity committed against him and sought to officially rehabilitate him as a martyr.[23] While this would finally give Montrose the decent death he had previously been denied, his funeral also served a broader political purpose. The Restoration gave rise to a wave of Royalist sentiment in which the themes of loyalty and sacrifice were carefully woven into the fabric of the regime.[24] Because of its political currency, Montrose's story was told and retold by Scottish Royalists into the eighteenth century and beyond.[25] In 1661, his courage in fighting for the King's cause and his defiance in the face of death was rewarded by a lavish funeral, the like of which had not been seen in Scotland since the coronation of Charles I in Edinburgh in 1633.

Montrose's attainment for treason in 1644 was intended to attach shame to his family's name and contribute to his social death in the eyes of the Covenant. However, in 1661, this social death was undone. Those "nearest in blood" to Montrose, including members of the Graham and Napier families, became a focal part of the funeral proceedings with one contemporary pamphlet commenting that the event marked a restoration of the good name of the Graham family.[26] On 7 January 1661, the funeral procession made its way through Edinburgh to the sound of drums, trumpets and the firing of cannons, to collect Montrose's torso from the Borough Muir. It was disinterred from under the gallows and carried under a velvet canopy to the Old Tolbooth where his head was taken down by members of his family, before the procession continued to Holyrood Abbey. He was placed in a coffin where he lay in state until the funeral was held in St Giles on 11 May, followed by a large banquet in his honour. During his invasion of Scotland, Oliver Cromwell had supposedly ordered Montrose's displayed limbs to be taken down. The arm sent to Aberdeen was interred in the vault of fellow Royalist George Huntly, 2nd Marquis of Gordon, who had been beheaded in 1649. In 1661, it was raised up and put in a velvet-covered box and carried by a procession of over 500 people through the city.[27] The celebration in Aberdeen was an important milestone in Montrose's public rehabilitation, as during his campaigns in the 1640s, he had attacked and plundered the city.

The funeral was conducted at the King's expense and was directed by Sir Alexander Durham who, as the Lyon King of Arms, was responsible for overseeing state ceremonies in Scotland. Durham's accounts show that he distributed, at least, the enormous sum of £802 sterling for Montrose's funeral. This lavish expenditure was more than mere remorse on the part of the monarch for a fallen cavalier. If we examine the great number of nobles and gentry who were present for the whole spectacle, it becomes clear that it brought together Montrose's friends and foes.[28] This demonstrates that the Restoration regime intended the spectacle to act as a vehicle to propagate the value of loyalty and to show its strength after a generation of civil wars. Following the funeral, Montrose's remains were buried in the cathedral of St Giles, in the vault of his grandfather, a previous Viceroy of Scotland.[29] Although the funeral and burial were intended to provide Montrose with an honourable death, not all of his body found its final resting place in the vault in St Giles. The next section looks at the separate journeys of one of Montrose's arms and his heart in order to highlight the beliefs that were attached to them and to demonstrate the continued agency of his body.

"A Most Convincing Form of Testimony"

There were various ways to think about, and be affected by, dead bodies in seventeenth-century Britain. These included: debates about when a person was medically dead; debates about the religious importance, or power of corpses; and beliefs about the potency of the dead and the healing properties of certain body parts.[30] Popular ideas about dead bodies were frequently noted at public executions, where people showed a desire to possess mementos, such as the blood, hair, clothing, and personal possessions of the executed person. Indeed, at the execution of Charles I, the monarch gave friends pieces of his clothing as relics, while after his death, his silk shirt and gloves became coveted curios. As was customary, the silk stockings worn by Montrose for his execution were claimed by the executioner. He had taken care not to cut them when severing the limbs, and after the event, they were purchased by Montrose's niece Elizabeth Erskine, Lady Napier. In 1856, a descendent of Lady Napier mentioned that the family was still in possession of the stockings, along with other relics of Montrose.[31] While this was an example of the repatriation of Montrose's possessions and memory by his family, Royalists were also concerned to "re-member" his body by tracing down his missing arm and heart.

It appears that the left arm sent to Aberdeen was the only one of the four distributed limbs to be collected in 1661. The right arm sent to Dundee to be nailed up above the principal town gate was subsequently carried to England by a Cromwellian officer named Pickering.[32] When one of Pickering's descendents left England for Spain in 1704, he placed the arm into Ralph Thoresby's antiquarian collection in Leeds. Upon Thoresby's death in 1725, the arm was purchased by Thomas Graham of Woodhall in Yorkshire.[33] It remained in the Graham family for decades and one his descendents, John Graham, wrote about the arm in 1752, stating its journey thus far and attesting to its authenticity. By 1834, Mr. C. Reeves of Woodhall, perhaps a descendent of the Graham family of Yorkshire, was in possession of the arm, and he provided details about its current condition. It was a mummified limb, he said, that had been cut off at the elbow and was in an excellent state of preservation.[34] In 1891, the arm was purchased by Mr. J.W. Morkill, along with a written statement of authenticity, and in 1925—the same year Charles' silk waistcoat was donated to the Museum of London—Morkill attempted to sell it at Sotheby's auction house. At the time, one newspaper stated that the arm was more than a gruesome relic because it offered a very definitive indication of the character of Montrose: "for understanding eyes it is an historical document, in addition to being a relic coveted by all who have fallen under the spell of a very gallant gentleman."[35] However, Morkill's notice of sale caused a public outcry and he withdrew the arm from Sotheby's. It was not mentioned again until 1932 when it was left to Morkill's son, Mr. Alan Greenwood Morkill, in his will.[36] After 1932, the arm disappeared from the historical record. Yet, despite the uncertainty of its final destination, the journey of the arm across almost three centuries demonstrates that people considered it to be a powerful curio as it was a tactile memento of the great Montrose worth possessing, and because the stories about it generated a sense of authenticity.

Following the execution and dismemberment, Montrose's torso was buried at the Borough Muir with the gallows used to hang him. However, when his body was disinterred in 1661, it was discovered that the chest had been broken open and heart removed. This had been done on the orders of his niece, Lady Napier. The post-mortem journey of Montrose's heart can be traced through a letter written by Alexander Johnson, a descendant of Lady Napier, to his daughter in the early nineteenth century.

According to Johnson, Lady Napier had the heart embalmed and enclosed in a case made from Montrose's sword. It was then placed into a gold filigree box that had been gifted to the Napier family from a Doge in Venice, and then placed inside a silver urn. She sent the heart to the Netherlands, where Montrose's son was in exile. After this, the heart was apparently lost for some time until a friend of the fifth Lord Napier recognised the gold box in a collection of curiosities in the Netherlands and purchased it for him. The Napier's had him sign a certificate attesting to its authenticity and the circumstances by which he had acquired it and Johnson wrote that it was then taken back to the Napier ancestral home at Merchiston. Johnson's grandfather often told the story of Montrose's heart to his mother and when he died he had left the heart to her. As a child, Johnson travelled to India with his mother and father, who was an officer in the East India Company, but on the voyage their ship had been attacked and the gold box was damaged. In India, his mother had it repaired by a goldsmith and also had another silver urn made with an engraving, in the two most common languages of the southern peninsula of India, telling the story of Montrose. The Johnsons displayed the urn in their home in Madura and, because of his mother's care for it, the locals believed it to be a talisman with the power to protect the bearer in battle. Owing to this superstition, it was stolen and sold to a powerful chief. But this was not the end of the heart's journey.

In his letter, Johnson recalled how he was often sent to hunt with local chiefs in order to learn more about their culture, and on one trip he earned the praise of a particular chief for an act of bravery. In a remarkable twist of fate, this was the chief who had purchased Montrose's heart, without knowing it had been stolen, and he agreed to return it to the family. (Interestingly, some years later, the chief was executed for his part in a rebellion, but before this, he told his attendants the story of Montrose and asked them to preserve his heart in the same way.) Montrose's heart then returned to Europe with the family in 1792, but during their journey through France, they found out that the revolutionary government was confiscating gold and silver. The Johnsons therefore entrusted the gold box into the safe keeping of an English woman in Boulogne named Knowles. When Knowles died soon after, the family were unable to trace the heart.[37] But still, the trail did not go cold.

In 1931, Captain H. Stuart Wheatley-Crowe, the president of the Royal Stuart Society, led an investigation into the missing heart. Wheatley-Crowe had in his possession an embalmed heart that was believed to have been brought to England from France during the Revolution by the ancestors of the Perkins family who claimed it was the heart of Montrose. He had a medical examination carried out upon the heart that found it to be approximately 300 years old, but could find no other definitive proof of its authenticity.[38] In 1951, Wheatley-Crowe sent the heart to Canada to a person he believed had a claim to the relic, a Mrs. Maisie Armitage-Moore.[39] Another turn of events came in 2012 when the largest ever collection of memorabilia marking the life of James Graham was exhibited in the Montrose Museum, a museum that opened in the town of Montrose in 1842, to mark the 400-year anniversary of his birth. The exhibition included paintings, documents, weapons, and a heart believed to be that of Montrose himself. The museum's curator acknowledged that there were two recorded accounts of different hearts believed to belong to Montrose and they had located one. However, it is unclear if this was the same heart that had been sent to Canada in 1951.[40] Despite the lack of proof of its authenticity, the heart was placed alongside other artefacts definitively related to Montrose, and this perhaps is a suitable final destination.

CONCLUSION

While awaiting his execution in the condemned cell of the Old Tolbooth, Montrose remarked to the guard "even after I am dead I will be continually present…and become more formidable to them [the Committee of Estates] than while I was alive".[41] Despite making this statement, Montrose could not have foreseen how both his body and his legacy would be utilised by both the Covenanting and the Royalist causes to propagate entirely different values. His three-day execution spectacle was replete with the hallmarks attached to the punishment for offences against the state, from the ignominious public procession to the multiple stages of the execution itself. Furthermore, the displaying of his corpse to indefinitely mark out his criminality was intended to prolong his legal death beyond the extinction of life. However, in conducting a public funeral, the Restoration regime changed Montrose's identity from that of an executed traitor to that of a murdered martyr and reconciled him religiously and legally.

We can draw parallels between the posthumous treatment of Montrose and other influential corpses from the Civil War, Interregnum and Restoration period. Following his execution for treason in 1681, the remains of Oliver Plunkett, the late Roman Catholic Archbishop of Armagh and Primate of All Ireland, were exhumed in 1683 and went on a journey of spiritual rehabilitation across Europe before he was eventually canonised in 1975. In contrast, upon his death in 1658, Oliver Cromwell, the late Lord Protector of the Commonwealth, received an elaborate state funeral which was intended to serve as a reinforcement of the Protectoral regime. However, this did not to mark his final resting place or his final legal death. By order of the Restoration regime, he was posthumously convicted and executed as a regicide with his spiked head on top of Westminster Hall serving as a reminder of the reward for treason.[42] The chronology of Cromwell's multiple deaths presents an almost reverse pattern to those of Montrose who suffered an ignominious execution in 1650, but received a lavish funeral in 1661 to mark his official death at the same time as his rehabilitation in the public memory.

Will Montrose ever die? This chapter has shown that, even after the honourable reparation afforded to Montrose in 1661, he was not, and perhaps is not yet, truly dead. Some of his body parts, once the dismembered remains of a traitor, were refashioned into coveted relics and instead of marking out his criminality, they attested to his gallantry and loyalty to the king. Spanning four centuries, the journeys of Montrose's arm and heart drew forth beliefs about body parts as signs of punishment, curious relics, icons of political memory, and curated exhibits. For the Covenant, Montrose's execution in 1650 marked the end of his life, but for Royalists it was the honourable funeral of 1661 that marked his legal death and repatriation into a political community. Montrose remains an iconic figure in Scottish history; indeed, the First Marquis of Montrose Society was founded in 1995 to promote his name and memory. This contemporary relevance, alongside the mobility and multiple meanings attached to his body parts, make it unlikely that Montrose's post-mortem journeys are over yet.

NOTES

1. Thomas W. Laqueur, "The Deep Time of the Dead", *Social Research*, 78 (2011), p. 802.
2. Mark Napier, *Memoirs of the Marquis of Montrose*, Vol. 2 (Edinburgh, 1856), p. 826.
3. *Yorkshire Post and Leeds Intelligencer*, 24 July 1925.
4. David Stevenson, *The Scottish Revolution 1637–1644* (Edinburgh, 2003), pp. 42–45.
5. John Buchan, *Montrose* (London, 1928), p. 107.
6. Robert Woodrow, *The History of the Sufferings of the Church of Scotland from the Restoration to the Revolution*, Vol. 1 (Glasgow, 1832), pp. 88–89.
7. Buchan, *Montrose*, p. 384.
8. Robert Chambers, *History of the Rebellions in Scotland 1638 till 1660*, Vol. 1 (Edinburgh, 1828), p. 11.
9. Buchan, *Montrose*; Ronald Williams, *Montrose: Cavalier in Mourning* (London, 1975).
10. Catriona M.M. MacDonald, "Montrose and Modern Memory: The Literary Afterlife of the First Marquis of Montrose", *Scottish Literary Review*, 6 (2014), pp. 4–7.
11. Chambers, *History*, 1, p. 218.
12. Buchan, *Montrose*, p. 365.
13. Napier, *Memoirs*, 2, p. 798.
14. Robert Chambers, *History of the Rebellions in Scotland 1638 till 1660*, Vol. 2 (Edinburgh, 1828), p. 231.
15. Buchan, *Montrose*, p. 374.
16. Daniel Szechi, "The Jacobite Theatre of Death". In Eveline Cruickshanks and Jeremy Black eds., *The Jacobite Challenge* (Edinburgh, 1988), p. 59.
17. *A Relation of the Execution of James Graham, Late Marquesse of Montrose* (London, 1650), p. 3.
18. *The Scots Remonstrance or Declaration Concerning the Restoring their Declared King to His Just Rights* (London, 1650), p. 2.
19. Napier, *Memoirs*, 2, p. 802.
20. Ibid., p. 809.
21. *The Substance of the Declaration of the Declared King of Scots upon the Death of the Marques of Montrosse* (London, 1650), p. 4.

22. George Wishart, *Memoirs of the Most Renowned James Graham, Marquis of Montrose* (Edinburgh, 1819), pp. 402–406.
23. Napier, *Memoirs*, 2, p. 826.
24. Bruce Lenman, *The Jacobite Risings in Britain 1689-1746* (Aberdeen, 1995), p. 20.
25. Keith M. Brown, "Courtiers and Cavaliers Service, Anglicisation and Loyalty among the Royalist Nobility". In John Morrill ed. *The Scottish National Covenant in its British Context 1638-51* (Edinburgh, 1990), p. 156.
26. *A Relation of the True Funerals of the Great Lord Marquesse of Montrose* (Edinburgh, 1661), p. 11.
27. Buchan, *Montrose*, p. 378.
28. Lenman, *Jacobite Risings*, pp. 21–22.
29. Napier, *Memoirs*, 2, p. 810.
30. Sarah Tarlow, *Ritual, Belief and the Dead in Early Modern Britain and Ireland* (Cambridge, 2011); Owen Davies and Francesca Matteoni, "'A Virtue Beyond All Measure': The Hanged Man's Hand, Gallows Tradition and Healing in Eighteenth- and Nineteenth-Century England", *Social History of Medicine*, 28:4 (2015), pp. 686–705.
31. Napier, *Memoirs*, 2, p. 810.
32. *Falkirk Herald*, 29 March 1905.
33. *Aberdeen Free Press*, 26 December 1891.
34. *Yorkshire Gazette*, 27 September 1834.
35. *Yorkshire Post and Leeds Intelligencer*, 24 July 1925.
36. *Yorkshire Post and Leeds Intelligencer*, 1 September 1932.
37. Napier, *Memoirs*, 2, pp. 819–824.
38. *Aberdeen Journal*, 6 March 1931.
39. *Hartlepool Northern Daily Mail*, 31 December 1951.
40. "Heart among Exhibits in Marquis of Montrose Collection". *BBC News* (accessed at http://www.bbc.co.uk/news/uk-scotland-tayside-central-19165945, dated 8 August 2012).
41. Wishart, *Memoirs*, p. 393.
42. Note that, following its removal from Westminster Hall, Cromwell's head became an object of curiosity and debate over its authenticity before it was eventually interred at Sidney Sussex College, Cambridge, in 1960. For more information on the respective journeys of the heads of Cromwell and Plunkett see Sarah Tarlow, "Cromwell and Plunkett: Two Early Modern Heads called Oliver". In James Kelly and Mary Ann Lyons eds., *Death and Dying in Ireland, Britain, and Europe: Historical Perspectives* (Dublin, 2013), pp. 59–76.

Author Biography

Rachel Bennett was recently awarded her Ph.D. in history by the University of Leicester and is now a Postdoctoral Fellow at the Centre for the History of Medicine at the University of Warwick. Her research interests are focused upon crime and punishment in Britain between the seventeenth and the nineteenth centuries, notably the history of capital punishment and the post-mortem punishment of the criminal corpse.

Open Access This chapter is licensed under the terms of the Creative Commons Attribution 4.0 International License (http://creativecommons.org/licenses/by/4.0/), which permits use, sharing, adaptation, distribution and reproduction in any medium or format, as long as you give appropriate credit to the original author(s) and the source, provide a link to the Creative Commons license and indicate if changes were made.

The images or other third party material in this chapter are included in the chapter's Creative Commons license, unless indicated otherwise in a credit line to the material. If material is not included in the chapter's Creative Commons license and your intended use is not permitted by statutory regulation or exceeds the permitted use, you will need to obtain permission directly from the copyright holder.

CHAPTER 4

Overcoming Death: Conserving the Body in Nineteenth-Century Belgium

Veronique Deblon and Kaat Wils

The start of the nineteenth century coincided with the development of a new aesthetics of death. Funerary rites became more elaborate; cemeteries and tombstones were increasingly adorned, and corpses were embellished before their burial. At the centre of this beautification of death movement was a new individualised and "sentimentalised" relationship with the dead.[1] Paying respect and tribute to the personhood of the deceased became an important aspect of mourning and funeral culture. At the same time, a more romanticised view of the afterlife was portrayed with an emphasis on the promise of reunion in life after death.[2] In European funerary culture, the "death as sleep" metaphor became a popular representation of death as it allowed mourners to separate the idea of death from a final state of being and corresponded with the Christian narrative of resurrection. In this narrative, the transition from life to

---

V. Deblon (✉) · K. Wils
University of Leuven, Leuven, Belgium
e-mail: veronique.deblon@kuleuven.be

K. Wils
e-mail: kaat.wils@kuleuven.be

death allowed the soul to fall asleep, only to awaken for the final judgement.[3] As Sarah Tarlow indicates, the comparison between sleep and death allowed for a "figurative understanding" of death (as the deceased were granted an eternal rest in peace).[4] The death as sleep narrative was accordingly part of the commemorative culture of Christian worship and it was materialised in tombstones and grave inscriptions.[5]

The grave as a peaceful place of sleep stands in sharp contrast with the reality and corporeality of death—of decomposition and putrefaction. The process of decay is inevitable in the body once organs stop functioning and bacteria and enzymes start decomposing bodily tissue. In the nineteenth century, new conserving procedures for the dead body were developed in response to a growing fear about the decaying corpses of the dead. New embalming techniques seemed to materialise the death as sleep metaphor: the preserved corpse appeared to be sleeping. These conservation methods were developed in the anatomical theatre, where corpses were preserved for scientific research or educational purposes. Even though embalmed corpses destined for burial and anatomical preparations were displayed in a different context, they both represented the corpse lingering between life (or sleep) and death. Moreover, defining death as sleep offered anatomists a visual language to connect their practices to contemporary funeral culture. By looking at the use, display, and popularisation of conservation procedures for corpses, we will show how death was overcome by preserving the body and suggesting a state of sleep in both medical and funeral culture. The display of sleeping corpses further aligned medical practice to Catholic death rites and an emerging conspicuous funeral culture.

This chapter presents a case study on nineteenth-century Belgium, where anatomists explicitly connected their practices to a funeral culture by preserving and displaying dead bodies as if they were still alive and seemed to be merely sleeping. The death as sleep metaphor received a new application in these treatments of the corpse. In doing so, anatomists testified to a sentimentalised relationship with individual corpses. At the same time, they helped shape nineteenth-century death culture by presenting their conserved corpses as if they were sleeping.

The practice of conserving corpses was not entirely new in Western Europe in the nineteenth century. This first part of the chapter looks at an early modern method for conserving bodies used by the Dutch anatomist Frederik Ruysch (1638–1731). Ruysch's work (re)gained renown in the nineteenth century when interest in the conservation and beautification of corpses was high. Inspired by Ruysch, the Belgian anatomist Adolphe Burggraeve (1806–1902) preserved corpses in such a way that they appeared to be alive and seemed to be sleeping peacefully. Focusing on the death as sleep metaphor, we examine how Burggraeve's "sleeping corpses" became familiar representations of death in an era in which embalming was popularised in Belgium. The final part of the chapter will look at how anatomical preparations impacted on death cultures and people's perceptions of what death looked like.

Conserving the Dead Body: The Preparations of Frederik Ruysch

For medieval physicians and churchmen, the conserving and collecting of body parts was a routine practice that formed an important aspect of how medical knowledge was passed on and how Christianity was practised. The first medical collections contained several osteological remains such as skeletons, while in churches different bones and skulls of saints were preserved. These collections of anatomical specimens and holy relics were designed to be displayed, and by the Renaissance period, they became increasingly adorned and made into commodities.[6] However, in the seventeenth century, anatomists working on (lay) bodies developed techniques to conserve tissues by macerating and storing body parts in alcohol-filled jars. This meant that they could create new spectacles of death, beyond the dry skeletons that originally filled their cabinets.

Frederik Ruysch showed particular interest in the conservation of corpses, both as embalmed and anatomical preparations. Ruysch was particularly known for his skeletons of foetuses that were preserved and often mounted in a "landscape" of human body parts, such as injected arteries or kidney stones. The lifelike skeletons—that were placed in an upright position—held different vanitas symbols, such as feathers, pearls, and handkerchiefs. The inscriptions that accompanied the anatomical preparations reminded visitors of the frailty of life. For example, labels of the tableaux read: "Death spares no man, not even the defenceless infant", or "What is life? A transient smoke and a fragile bubble".[7] The tableaux of foetal skeletons were intended to present spectators with a moral message on death as the inevitable fate of man and served as memento mori.

Ruysch also developed a conservation technique for the preservation of human tissue by injecting them with substances such as (coloured) wax or mercury and storing them in alcohol. Though Ruysch was not the first to employ this method, he developed the skill to create specimens that mimicked the body's natural state.[8] The uncanny character of the preparations came from their lifelike appearance: contemporary visitors to Ruysch's collection emphasised the "rosy complexions" of the bottled babies, which made it appear as if "they had never died".[9] His injection technique not only restored the suppleness of the body in its living state, it also allowed him to map fine anatomical structures such as the capillaries.

Ruysch's lifelike specimens also formed a tribute to the infinite power of a divine Creator. His cabinet was recommended as a place where visitors stood "face to face with manifestations of the creation" and could observe the ingenious structure and functioning of the human body as it was designed by the Creator.[10] In his study on the preparations of Ruysch, Van de Roemer emphasises that the attributes of the specimens, such as embroidered textile, strengthened the religious reflections of visitors to the anatomical cabinet. The delicate lace that covered the anatomical preparations offered a visual association between the texture of textile (made by man) and the texture of the body (as created by a divine power).[11] Moreover, the lifelike appearance of the preparations supported the emotional connection between dead body parts and their spectators.

Given this link between the display of bodies and musings on mortality and Creation in early modern collections, anatomical preparations could be seen to have similar functions to holy relics.[12] Anatomical preparations offered mourners a "material and emotional link to the deceased" and "represent[ed] the living" in the same way that relics did.[13] While relics had a power associated with their origin in the bodies of saints and narratives of martyrdom, why did Ruysch's preparations need to be beautiful in order to inspire people to reflect on death? The artistic embellishing of preparations helped to deal with people's disgust at the corpse; the anatomical artistry consciously referenced the divine work of God, a juxtaposition that gave anatomists like Ruysch a spectacular power.[14] Marieke Hendriksen further argues that the "elegancy" of anatomical preparations in the early modern period was a necessary means by which anatomical knowledge was transmitted. The senses of touch, sight, and smell might overpower those who approached the (unpreserved) corpse to learn, yet the beauty of the preparations demonstrated the beauty of the anatomical body. In turn, the perfection of dissected or injected body parts generated anatomical knowledge.[15] Though the taste for elegance in seventeenth-century anatomical preparations soon waned, Ruysch's work was celebrated until well into the nineteenth century precisely because he was able to eliminate "the disgust inspired by the cadaver".[16]

Death in Nineteenth-Century Belgium

In the early modern period, representations of the decaying body served as memento mori—a reminder of one's own mortality—but by the nineteenth century, the thought of bodily dissolution came to be seen as distressing. Nineteenth-century death culture aimed to conceal the "reality of death and decay".[17] This was because, from the late eighteenth century onwards, there was a transformation in how people mourned. An individualised relationship with the dead emerged and burial became increasingly aestheticised as part of the wish to pay tribute and respect to the personhood of the deceased.[18] This created a new sensitivity about the decay of the corpse, because it threatened to affect people's emotional relationship with the dead, and raised fears about the medical dangers of rotting bodies.

In nineteenth-century funeral rituals, more attention was paid to the bodily integrity of the deceased. Amid the cultural shift towards an individual relationship with the dead, anatomists and medical students at European universities developed new and more efficient ways to conserve the corpse for use in their studies. Their desire to halt decay did not exclude their desire to beautify the corpse; rather, these new conservation methods overlapped with the aesthetic methods that upper-class mourners were using in their funerary practices.[19]

In France, new embalming techniques were popularised and commercialised by medical men in the first half of the nineteenth century. A similar shift occurred in America where new conservation techniques were marketed by funeral directors in the emerging funeral industry.[20] In Belgium, funeral rites and ceremonies were exclusively in the hands of the Catholic Church in the first half of the nineteenth century. Religion played a crucial part in the moments leading up to death and in the organisation of funerals. Confession, absolution, and prayer accompanied the final hours of a dying person.[21] In Catholicism, the final sacraments allowed the soul to live on in the afterlife and supported the idea of heavenly immortality.

After death, the laying out of the corpse was part of a dignified farewell. This particular task was usually performed by the sisters of a Catholic order and was often reserved for the highest echelons of society. For example, the bodies of bishops were washed, covered with odorous spices and laid out after death. Displaying the corpse was seen as a mark of respect and appreciation for the deceased. Care for the dead body was an expression of faith in resurrection and the immortality of the soul.[22] Cherishing the corpse would provide that dead body with the opportunity to resuscitate and reunite with the soul in the afterlife. Physicians who wanted to promote their embalming techniques capitalised on this belief: conserving the corpse equalled care for the corpse and guaranteed its physical integrity.

In the homes of the destitute, the laying out of corpses was more of a necessity than an aspect of the funeral rite. Due to the lack of municipal morgues, pauper corpses often stayed in people's one-bedroom houses in attendance of a funeral.[23] As the embalmment of corpses in nineteenth-century Belgium was exclusively reserved for royals, nobility, and the bourgeoisie, the preservation of pauper corpses in the domestic sphere was often problematic.[24] Decaying pauper corpses were increasingly seen as a medical hazard. The time between the moment of death and the actual funeral was therefore strictly limited. Yet, the presence of the beloved deceased in the domestic sphere also contributed to the image of the "corpse at peace".[25] In her study on the expression of grief among the working classes in nineteenth-century Britain, Julie-Marie Strange notes that the "relaxation of the facial muscles" and the similarities between "shroud and nightdress" led spectators to associate death and sleep.[26]

The post-mortem fate of paupers who died in the hospital was radically different from those who passed away at home. Patients' corpses were transferred to the anatomical theatre and dissected when the family was unable to cover the funeral costs. The putrefying corpses in the anatomical theatre evoked repugnance among the people in the neighborhood who often complained about bad odours. The practices of anatomists therefore enjoyed a bad reputation among a non-medical audience. Anatomists tried to disconnect the corpses from the realities of death by embalming and conserving their source material.

BETWEEN LIFE AND DEATH: THE PREPARATIONS OF ADOLPHE BURGGRAEVE

While Ruysch's anatomical preparations had been excessively adorned with textile or attributes in the early modern period, by the end of the eighteenth century, a more realist, "representational" and modest style was developed for anatomical representations of the body.[27] Elaborate displays of decorated body parts went out of fashion. As anatomy moved away from providing spectacles of death for lay audiences, anatomists developed more sober representations of the anatomical body.[28] However, the desire to overcome the disgust and horror of the corpse remained an important aspect of the work of anatomical collectors into the nineteenth century.[29] This removal of disgust chimed in with contemporary mourning cultures and provided a new context for the display of anatomical preparations.

In Belgium, the creation and institutionalisation of anatomical collections was an important aspect of the development of a national science after its independence in 1830. Anatomists at the new universities put great effort in the establishment of anatomical museums. At the University of Ghent, Adolphe Burggraeve and his aide Edouard Meulewaeter developed several conserving techniques to ensure the expansion of the institution's anatomical collection in the 1830s. Together they created mercury-injected specimens, injected bone preparations, and injected preparations of the mucous membrane next to larger wet specimens of diseased organs or foetuses. Their preparations were lauded for their natural appearance and delicacy, and were seen as the ultimate proof of Burggraeve's craftsmanship. Though elegance went out of style, the preparations housed in the anatomical museum at the University of Ghent show that anatomists were still concerned about creating the most lifelike preparations possible.

Burggraeve searched for years for a preservation method that could equal the results of Ruysch. Inspired by an encounter with some lifelike preparations in the Netherlands, Burggraeve devised a procedure that he felt matched that of Ruysch. With the assistance of his aide Meulewaeter, Burggraeve found a way to inject corpses with coloured gelatine, and then conserved the preparations in soured alcohol. They added over a 1000 specimens to the University's anatomical cabinet, but only a few preparations from the original collection have survived. One of them is a newborn child dressed in a white christening gown, floating in a transparent glass jar (Fig. 4.1). This preparation received considerable attention in the nineteenth-century medical press; its natural skin tone and the rosiness of the cheeks and lips were particularly lauded. The velutinous quality of the skin was said to give the preserved body the "transparence of life" and suggested the bodies were at peace.[30] A preparation of a half-dissected girl displays a similar serenity, but paradoxically it also incorporates the violence of a dissection (Fig. 4.2). On the one hand, the girl on display is clearly dead; her head is cut in half to show the matter within. On the other hand, she appears to be sleeping peacefully. When, in 1837, Burggraeve presented some of his preparations to a medical audience, their lifelike appearance caused many to recall the work of Ruysch, an association Burggraeve was keen to highlight. Indeed, like Ruysch, Burggraeve adorned his preparations with textile decorations and kept his conservation method a secret.[31]

The purpose of the textile decorations was twofold: Firstly, it was meant to cover up mutilated body parts, as was its function in the seventeenth century.[32] The autopsy scars on the body of the newborn infant, for example, were concealed by the christening dress. Secondly, the adherence of textile to body parts also had a symbolic function: the christening gown of the newborn infant emphasised the innocence of the child, but also confirmed the possibility that the child would live a peaceful afterlife.[33] In the nineteenth century, it was customary to bury babies in their baptismal clothing as proof that they had been baptised and could enter the Kingdom of Heaven.[34] By dressing the child in a white gown, Burggraeve aligned the preparation of the child with contemporary funeral practices.

As an anatomical preparation, the body of the newborn child hardly demonstrated anatomical structures or knowledge. However, it did visualise a new "death way" to visitors to the anatomical museum. The dress softened the harsh reality of death, made the corpse into an aesthetic object, and hid the scars inflicted by the post-mortem examination.In Burggraeve's preparation of the girl, the reality of the dissection was not hidden. Rather, the traces of the dissection and the carving of the scalpel were clearly visible on the body. Her trunk appears to have been detached violently from her chest, and Burggraeve highlighted her anatomical structures by means of coloured gelatin injections. The textile attached to the preparation, however, served the symbolic purpose of drawing out an association with sleep. Draped as if it were a pillow, the

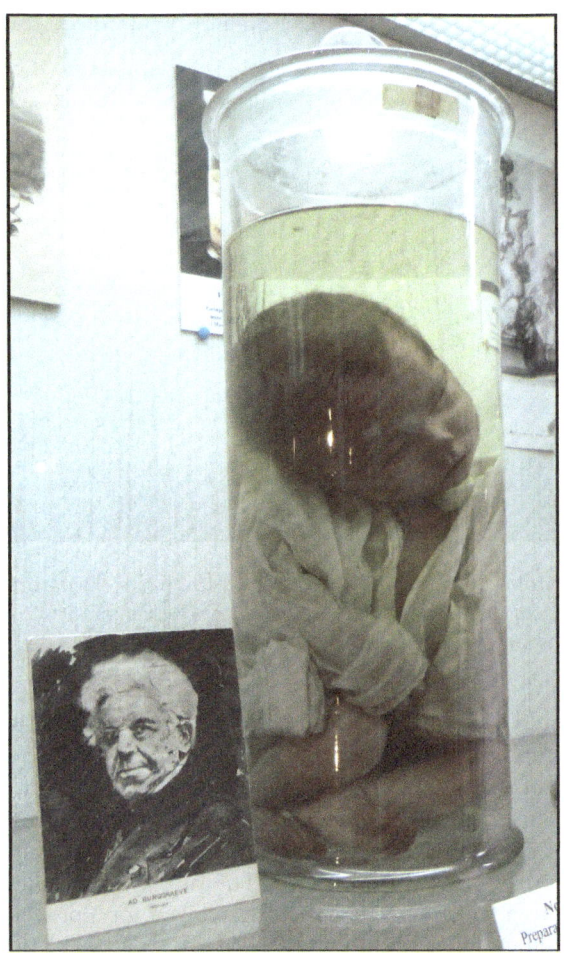

Fig. 4.1 Anatomical preparation of a newborn child (Museum for the History of Medicine, Ghent. Collection: University Museum, Ghent)

fabric seems to suggest that the deceased is resting and has found peace, despite the disintegrated state of her body.

Burggraeve's preparations not only caught the attention of his fellow medical professionals, but the results of his injection method were also noticed in the Belgian newspaper press where audiences read about how

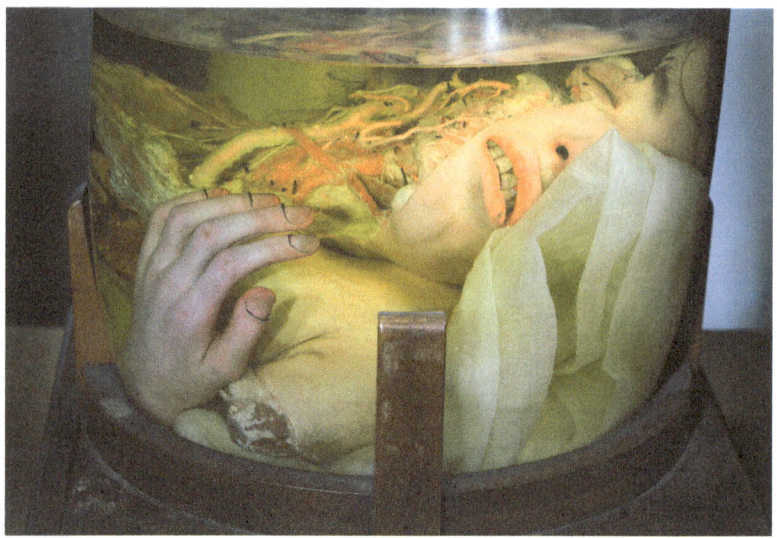

Fig. 4.2 Anatomical preparation of a girl with hand (Department of Basic Medical Sciences, Anatomy and Embryology Research Group. Collection: University Museum, Ghent. Photographer: Benn Deceuninck)

one could hold onto the bodies of the deceased. Burggraeve's invention was contextualised by referring to the human fascination with the dead body and our wish to overcome the "destruction" of life. In one newspaper article the "idea of death" was said to be "too dreadful for man", which explained, for the journalist, why many different cultures embalmed and conserved the dead.[35] The display of lifelike anatomical preparations to a lay audience, it is clear, allowed for a peaceful and consoling confrontation with death.

Embalmment Mania: Jean-Nicholas Gannal's Conservation Method

Anatomists continually sought for better ways to conserve putrefying corpses. In the anatomical theatre, the same corpse that served as an object of scientific enquiry one day could be the locus of fears about illness and disease from medical students the next day. A fear of rotting corpses also prevailed among the citizens of densely populated areas who saw the dead body as a breeding ground for diseases such as typhus and cholera.[36] In this context, anatomists intervened in medical and popular fears about the dead by developing ways to conserve bodies, preserve their physical integrity and extend their "liveliness". In the 1820s and 1830s, the French chemist Jean-Nicholas Gannal (1791–1852) gained critical acclaim for his procedure to embalm cadavers. After working as an apothecary in the army of Napoléon Bonaparte, Gannal grew interested in the conservation of anatomical preparations and small animals, and later created a method for embalming human corpses from his operating base in Paris. He did this by making a small incision in the neck, rinsing the arteries, and then injecting the corpse with arsenic. At first, only corpses destined for the anatomical theatre were treated this way, but later on he also developed a conservation method that became popular in the funeral industry.[37]

As news about Gannal's embalming procedure spread across Europe, hospitals and hygiene commissions became interested in applying his method for the conservation of corpses for dissection. The Medical Commission of the city of Brussels looked to Gannal's method as a possible solution to the health risks and insalubrity caused by the circulation of cadavers for anatomical courses.[38] Anatomists claimed that the embalming of corpses could halt the exhalation of "mephitic miasmas" and the emanation of "poisonous gas", which caused typhoid fever.[39] The teaching hospitals—where corpses were dissected—also reacted positively to Gannal's method. Embalming was considered an excellent solution to the problem of the high number of accidents and illnesses caused by *"le piqûre anatomique"*, or injuries caused by infection from dissection instruments.[40] In 1841, four years after Gannal had patented his injection method, Cécilien Simonart, the prosector of the Brussels anatomy department, approved Gannal's method for the conservation of anatomical preparations and for embalming procedures.[41]

Following the popularity of Gannal's conservation procedure, many scientists took it upon themselves to create new embalming methods. Research on new embalming procedures rocketed in the 1840s, which caused the *Gazette médicale belge*—a publication that regularly reported on new conserving techniques—to ask its readers "when will this embalmment mania stop?"[42]

The Application of Conserving Procedures

News on Gannal's embalming technique reached Belgium at the same time as Burggraeve displayed his preparations. Like Burggraeve's method, Gannal's injection technique received considerable attention in the newspaper press where it was also presented as a successor to Ruysch's procedure.[43] Burggraeve's anatomical preparations and Gannal's embalmed corpses were described in similar terms by journalists, who lauded the freshness of their conserved bodies that had all the "appearances of life".[44]

Both of these conserving procedures were developed behind the doors of the anatomical theatre, but it was Gannal who successfully tapped into a new market by attracting the interest of the funeral industry in France. In fact, Gannal claimed to have created two methods for the embalming of bodies. The first method was used in the anatomical theatre, whereas his second injection technique was used in funeral treatments. His commercial endeavours were so successful that Gannal patented his embalming method for funerals. Gannal's fame grew and Belgian newspapers eagerly reported on the eccentric embalmer from Paris who had transformed the way dead bodies appeared. As one journalist wrote: "we can say that the physiognomy of death has become nothing more than that of sleep and rest".[45] Gannal's procedure gained even more fame when it was allegedly used to conserve the already decaying body of Napoléon in 1840 after its exhumation.[46]

In 1845, the anatomist Charles Poelman (1815–1874) acquired Gannal's embalming patent for Belgium and received critical acclaim for it among his fellow physicians. Gannal's method had been celebrated for its low cost, allowing for a "popular application" of the embalming procedure.[47] Although Poelman did not use the method to advance medical education, the medical press in Belgium supported its application because it evoked the "interest and affection of families" who wished for their beloved ones to "remain after death the way they had been".[48] Another reason for the popularity of the Gannal method was that it respected the integrity of the body during the conservation process. In contrast to earlier embalming techniques, Poelman did not remove organs or the brain, thought of by Poelman as the instrument of the soul, "that which man made into man".[49] This respectful technical process supported the desire among upper-class Belgians for corpses to appear as if they were sleeping.[50]

Through his embalming, Poelman responded to, and redefined, funeral rituals in Belgium. He acknowledged the growing importance of the tomb as a "pious object of veneration that speaks to our soul" and place where the living and the dead could reunite in an "invisible communion".[51] Poelman urged that people considering a tomb should also consider embalming as a means to commemorate the dead, something he called a "sacred obligation".[52] As presented by Poelman, embalming was a consolation for the living and a means by which they could still maintain a relationship with the dead.

In the 1840s, just as Poelman promised to create a corpse that appeared to be sleeping, a Belgian pharmacist named Joseph Michiels promised to immortalise the corpse as a kind of living statue. Michiels experimented with several bodies in an effort to conserve them by covering them with copper, gold, or silver. His efforts were praised in the Belgian medical press as a method to conserve pathological specimens and as a form of embalming.[53] Michiels presented his preparations at one of the meetings of the Belgian Royal Academy of Science and Fine Arts in 1843, where an audience admired the "perfect representation of the traits of the face" in the preparations.[54] Though Michiels presented his conservation method in the form of anatomical preparations, the scientific community encouraged and discussed its application in a non-medical context: "families can place their parents in accessible galleries, as an alternative to enclosing them in dark tombs".[55] Instead of tracing the features of the face, as the death mask process did,[56] Michiels traced the entire body and transformed it into a sculpture, or a "statue for everyone", as one local newspaper reported.[57] Michiels's conservation method allowed families to hold on to the corpse, literally, and keep it at home as if it were a jewel or decoration. As the *Journal de Médecine de Bordeaux* noted: "it seems that a gilded corpse would do well in a comfortably furnished apartment".[58]

Overcoming Death: The Post-Mortem Subject

In the anatomical theatre, the corpse was fragmented and exposed for the purposes of medical education and spectacle, but in a funeral context, the integrity and respectful treatment of the corpse was paramount. Medical historians therefore connect the practice of anatomy with processes of objectifying the corpse.[59] By cutting, injecting, and displaying body parts, anatomists turned the corpse into an object that lacked personhood. However, the creators of anatomical preparations frequently elided this process by signposting the identity and personality of the deceased.

Take Burggraeve's anatomical preparations at the University of Ghent. The newborn infant was conserved in its entirety and wore a gown that symbolised youth and innocence. It also suggests a link with the family of the deceased infant. Baptism clothing was often recycled from the mother's wedding dress and was used to baptise several children from the same family. Burggraeve probably used the dress to suggest that the separation between the child and its family was only temporary. While this kind of preparation appealed to upper class Belgians, it was likely that the child was from the poor and destitute class that usually provided bodies for anatomists.

Burggraeve's preparation of the girl, meanwhile, made her recognisable and appealed to visitors who saw a person sleeping in a glass tomb. The addition of the hand to the preparation gave an extra dimension to the display of the body, emphasising the personality, femininity, and elegance of a body that was fragmented and placed in a jar. Like the textile, the hand also served a symbolic purpose, probably referring to the well-known preparations of children's hands by Ruysch and the Dutch anatomist Bernhard Siegfried Albinus (1697–1770). The excellence of the conserved hand also alluded to the manual skills of the anatomist.[60] Although both of these bodies were anonymous, visitors sensed the subjectivity of the people in the jars and this made death seem less unpleasant. As a journalist put it, "we can please ourselves by contemplating the sweet animation of life in the corpse".[61] Burggraeve's bodies can be defined as "post-mortem subjects", a category John Troyer uses to think about the emergence of preservation technologies that made "the dead body look more alive".[62] Burggraeve's preparations seem to hover somewhere between life and death, an uncanny situation that led one journalist to describe how "the tricked eye can still see the blood and life circulating".[63]

In the preparation of the girl, the presence and colour of the hand suggests life, but the state of the head points to the liminal position of the anatomised subject. Her head is literally positioned between life and death: on the one side she is peacefully sleeping; on the other side, her open skull makes the brain visible to visitors. In this, the preparation resembled nineteenth-century wax anatomical models in its form and in its connection with funerary sculptures through their incorporation of life and death in the body.⁶⁴ This disconnection from the realities of death echoed a wider cultural transformation in the status of the dead, embalmed body. By the nineteenth century, the scientific value of elegant or elaborately decorated anatomical preparations had diminished. However, Burggraeve's lifelike preparations had renewed value as icons that could diminish people's fear of corpses. The beauty of these preparations becomes "apparent in contrast to [their] destruction", disguising the presence of death and decay.⁶⁵

CONCLUSION

The work of nineteenth-century anatomists and embalmers impacted on how people were mourned and their dead bodies considered in Belgium. Preservation showed that death did not inevitably lead to decay, that death could be thought of as a sleeping. In the Belgian newspaper press, Gannal and Burggraeve's techniques were compared to the work of Ruysch, who had used a secret method to create his lifelike preparations. This chapter has argued that anatomists and embalmers materialised the death-as-sleep metaphor. Belgium shared with other nations a concern about the health issues associated with decaying bodies and anatomical practice. On an individual level, people desired an illusion that death was not the end, and that the personality of the deceased could be frozen in the corpse itself. Responding to this, anatomists developed methods for conservation in a medical context which received broader cultural applications. The displays discussed in this chapter demonstrate the journey of conserving and embalming practices from the anatomical theatre to the funeral parlour, from the world of medical education to the world of mourning practices. While anatomists could exert control over bodies and their decay through the use of preservatives and chemicals, visitors to anatomical cabinets could feel that they also exerted control over their own destiny by imagining that dead bodies looked like they were sleeping.

NOTES

1. Philippe Ariès, *The Hour of Our Death* (Harmondsworth, 1981), pp. 409–474.
2. Teresita Majewski and David Gaimster, *International Handbook of Historical Archaeology* (Berlin, 2009), p. 145.

3. Nas E. Boutammina, *La Mort: Approche Anthropologique et Eschatologique* (n.p., 2015); W.M. Spellman, *A Brief History of Death* (London, 2014), p. 125; Jozef Lamberts, "De Rooms-Katholieke Uitvaartliturgie". In Lambert Leijssen et al. eds., *Dood en Begrafenis* (Leuven, 2007), pp. 119–131.
4. Sarah Tarlow, "Belief and the Archaeology of Death". In Liv Nilsson Stutz and Sarah Tarlow eds., *The Oxford Handbook of the Archaeology of Death and Burial* (Oxford, 2013), p. 620.
5. Sarah Tarlow, *Bereavement and Commemoration: An Archaeology of Mortality* (New York, 1999); Tarald Rasmussen and Jon Øygarden Flæten eds., *Preparing for Death, Remembering the Dead* (Göttingen, 2015), p. 35.
6. Myriam Nafte, "Institutional Bodies: Spatial Agency and the Dead", *History and Anthropology*, 26:2 (2015), pp. 206–233; Ken Jeremiah, *Christian Mummification: An Interpretative History of the Preservation of Saints, Martyrs and Others* (Jefferson, North Carolina, 2012).
7. Christine Quigley, *Skulls and Skeletons: Human Bone Collections and Accumulations* (Jefferson, North Carolina and London, 2001), p. 184.
8. Dániel Margócsy, *Commercial Visions: Science, Trade, and Visual Culture in the Dutch Golden Age* (Chicago, 2014), p. 111.
9. Luuc Kooijmans, *Death Defied: The Anatomy Lessons of Frederik Ruysch* (Leiden and Boston, 2011), p. 315.
10. Gijsbert M. van de Roemer, "From *Vanitas* to Veneration: The Embellishments in the Anatomical Cabinet of Frederik Ruysch", *Journal of the History of Collections*, 22:2 (2010), p. 178.
11. Ibid., p. 182.
12. Rina Knoeff, "Ballpool Anatomy: On the Public Veneration of Anatomical Relics". In Robert Zwijnenberg and Rina Knoeff eds., *The Fate of Anatomical Collections* (Farnham, 2015), pp. 279–292.
13. Ibid., p. 288.
14. Van de Roemer, "From *Vanitas* to Veneration", p. 181.
15. Marieke M.A. Hendriksen, *Elegant Anatomy: The Eighteenth-Century Leiden Anatomical Collections* (Leiden and Boston, 2014), pp. 11–12.
16. Adolphe Burggraeve, *Cours théorique et pratique d'anatomie: comprenant l'histoire de l'anatomie, l'ovologie, l'organogénésie et les monstruosités, l'anatomie des tissus et l'anatomie pathologique* (Ghent, 1840), p. 296.
17. George Nash, "Pomp and Circumstance. Archeology, Modernity and the Corporatisation of Death: Early Social and Political Victorian Attitudes towards Burial Practice". In Paul Graves-Brown ed., *Matter, Materiality, and Modern Culture* (London and New York, 2000), p. 114.
18. Michael Sappol, *A Traffic of Dead Bodies: Anatomy and Embodied Social Identity in Nineteenth-Century America* (Princeton, New Jersey, 2004), p. 18.

19. Anne Carol, *L'Embaumement. Une Passion Romantique: France, XIXe Siècle* (Seyssel, 2015), p. 117.
20. Gary Laderman, *Rest in Peace: A Cultural History of Death and the Funeral Home in Twentieth-Century America* (Oxford, 2003).
21. Patricia Jalland, *Death in the Victorian Family* (Oxford, 1996), p. 18.
22. *Précis Historiques. Bulletin des Missions Belges de la Compagnie de Jésus, Congo, Bengale, Ceylan* (Brussels, 1855), p. 359.
23. "Enterrement des Pauvres à Bruxelles", *Assemblée Générale des Catholiques en Belgique. Première session à Malines, 18–22 août 1863*, Vol. 2 (Brussels, 1864), pp. 356–357.
24. Christoph De Spiegeleer, "Sterven, Begraven en Herdenken van Koninklijke en Politieke Elites in België tussen 1830 en 1940. Een Culturele en Politieke Geschiedenis" (Unpublished doctoral dissertation, Vrije Universiteit Brussel, 2016), p. 142.
25. Julie-Marie Strange, *Death, Grief and Poverty in Britain, 1870–1914* (Cambridge, 2005), p. 77.
26. Ibid., p. 78.
27. Michael Sappol, *Dream Anatomy* (Bethesda, Maryland and Washington, D.C., 2006), pp. 25–34; Jonathan Simon, "The Theater of Anatomy: The Anatomical Preparations of Honore Fragonard", *Eighteenth-Century Studies*, 36:1 (2002), pp. 63–79; Hendriksen, *Elegant Anatomy*, pp. 178–181.
28. Simon, "The Theater of Anatomy", p. 75.
29. Elizabeth Hallam and Samuel Alberti, "Bodies in Museums". In Elizabeth Hallam and Samuel Alberti eds., *Medical Museums: Past, Present and Future* (London, 2013), p. 4.
30. Hippolyte Kluyskens et al., "Rapport sur les Préparations Anatomiques de M. le Professeur Burggraeve, lu dans la Séance de la Société de Médecine de Gand du 7 Novembre 1837", *Société de Médecine de Gand* (1837), p. 320.
31. Veronique Deblon and Pieter Huistra, "Het Geheim van de Anatoom: Adolphe Burggraeve en de Ontwikkeling van de Belgische Anatomie in de Negentiende Eeuw". In *Studium. Tijdschrift voor Wetenschaps-en Universiteitsgeschiedenis*, 9:4 (2017), pp. 202–216.
32. Marieke M.A. Hendriksen, "The Fabric of the Body", *Histoire, Médicine et Santé*, 5 (2014), pp. 21–32.
33. Kitty de Leeuw, *Kleding in Nederland, 1813–1920* (Hilversum, 1992), p. 431.
34. Lou Taylor, *Mourning Dress: A Costume and Social History* (London, 1983), pp. 137–138.
35. "Conservation des Cadavres", *L'Indépendant*, 26 August 1837.

36. Cindy Stelmackowich, "Bodies of Knowledge: The Nineteenth-Century Anatomical Atlas in the Spaces of Art and Science", *RACAR: Revue d'Art Canadienne / Canadian Art Review*, 33:1/2 (2008), pp. 75–86; Jonathan Strauss, *Human Remains: Medicine, Death, and Desire in Nineteenth-Century Paris* (New York, 2012), p. 95.
37. Carol, *L'Embaumement*, pp. 55–56.
38. H.J. Vandencorput, "Conservation des Cadavres par le Procédé de Mr. Gannal" (1840). Letter from the Medical Commission to the City Council, File 58 (General Affairs: Correspondence concerning the Gannal procedure for the conservation of cadavers in the amphitheatre), Archives Social Services Brussels.
39. Ibid.
40. André Uytterhoeven, "Avis au Conseil d'Administration des Hospices et Secours concernant le Procédé Gannal" (1840). File 58 (General Affairs: Correspondence concerning the Gannal procedure for the conservation of cadavers in the amphitheatre), Archives Social Services Brussels.
41. *L'Indépendance Belge*, 20 August 1841.
42. *Gazette Médicale Belge*, 28 April 1844, p. 76.
43. "De l'Embaumement et de la Mommification", *Le Messager de Gand*, 3 May 1837.
44. Kluyskens et al., "Rapport", p. 320.
45. "De l'Embaumement".
46. *Le Messager de Gand*, 24 August 1840.
47. "Embaumements", *Bulletin de la Société de Médecine de Gand*, 11 (1845), pp. 124–125.
48. Ibid.
49. Charles Poelman, *Notice sur les Embaumements: Procédé Gannal* (Ghent, 1845), p. 9.
50. Carol, *L'Embaumement*, p. 146.
51. Poelman, *Notice*, p. 3.
52. Ibid., p. 4.
53. "Conservation, Embaumement et Modelage en Relief des Cadavres et Pièces Anatomiques, au Moyen de la Galvanoplastie", *Gazette Médicale Belge* (1843), p. 141.
54. Claude-Louis Sommé, "Nouveau Procédé pour la Conservation des Corps", *Bulletin de l'Académie Royale des Sciences et Belles-Lettres de Bruxelles* (1843), pp. 23–24.
55. Ibid., p. 24.
56. Quigley, *Skulls and Skeletons*, p. 43.
57. "A Chacun sa Statue", *Le Messager de Gand*, 22 July 1843.
58. "Vicissitudes de l'Embaumement", *Journal de Médecine de Bordeaux* (1843), pp. 519–520.

59. Samuel J.M.M. Alberti, *Morbid Curiosities: Medical Museums in Nineteenth-Century Britain* (Oxford and New York, 2011), pp. 99–101.
60. Hendriksen, *Elegant Anatomy*, p. 84.
61. "Conservation, Embaumement".
62. John Troyer, "Embalmed Vision", *Mortality*, 12:1 (2007), p. 24.
63. "Conservation, Embaumement".
64. Elizabeth Hallam and Jenny Hockey, *Death, Memory and Material Culture* (Oxford and New York, 2001), p. 65.
65. Elisabeth Bronfen, *Over Her Dead Body: Death, Femininity and the Aesthetic* (Manchester, 1992), p. 5.

Authors' Biography

Veronique Deblon is a Ph.D. student at the University of Leuven. In 2013 she joined the project "Anatomy, Scientific Authority and the Visualized Body in Medicine and Culture" at the research group "Cultural History since 1750". Her research examines the changing styles and representations of the anatomical body in the first half of the nineteenth century.

Kaat Wils is professor in European cultural history and head of the "Cultural History since 1750" research group at the University of Leuven. Her primary research focus is on the modern history of the human and the biomedical sciences, gender history, and the history of education. She recently co-edited *Bodies Beyond Borders. Moving Anatomies, 1750–1950* (Leuven University Press, 2017).

Open Access This chapter is licensed under the terms of the Creative Commons Attribution 4.0 International License (http://creativecommons.org/licenses/by/4.0/), which permits use, sharing, adaptation, distribution and reproduction in any medium or format, as long as you give appropriate credit to the original author(s) and the source, provide a link to the Creative Commons license and indicate if changes were made.

The images or other third party material in this chapter are included in the chapter's Creative Commons license, unless indicated otherwise in a credit line to the material. If material is not included in the chapter's Creative Commons license and your intended use is not permitted by statutory regulation or exceeds the permitted use, you will need to obtain permission directly from the copyright holder.

CHAPTER 5

Premature Burial and the Undertakers

Brian Parsons

Although legislation to register deaths was introduced in England and Wales in the 1830s, it was not mandatory for a physician to examine a body after death. For many people, the absence of a final check for signs of life led to fears of premature burial. In the 1890s, a pressure group called the London Association for the Prevention of Premature Burial (LAPPB) was founded to highlight the issue and it campaigned for improvements in death certification along with the building of "waiting" mortuaries. Despite this move, the lack of any large-scale evidence that people were being buried alive meant that there was little interest in achieving greater changes in law or practice. Furthermore, by this stage, the context of the disposal of dead bodies was changing, a development evident in the increased responsibility undertakers were given for the treatment of the corpse. This might have been an opportunity for undertakers to promote themselves as verifiers of mortality, offering the bereaved peace of mind by testing for death. However, the tentative moves to provide this service failed to gain any legitimacy because undertakers were seen as usurping the role of medical professionals, traditionally regarded as the chief certifiers of death.

B. Parsons (✉)
London, UK
e-mail: bparsonsfstl@gmail.com

© The Author(s) 2017
S. McCorristine (Ed.), *Interdisciplinary Perspectives on Mortality and its Timings*, Palgrave Historical Studies in the Criminal Corpse and its Afterlife, DOI 10.1057/978-1-137-58328-4_5

Historical in its approach, this chapter explores the landscape of the disposal of the dead in Britain from the 1830s, focusing, in particular, on the tension caused by the absence of secure certification of the dead and the changing role of the undertaker.

Concerns about premature burial, or "vivisepulchure" as Behlmer terms it, have existed in Britain since the eighteenth century, if not longer.[1] This fear inspired many writers to speculate on the horrors of waking up in a coffin after interment: Edgar Allen Poe's "The Premature Burial" (1844) is one of most widely known takes on the subject. The possibility that a trance-like state could be mistaken for death stimulated people to invent "safety" coffins, many kitted out with warning devices that could be activated by the supposedly deceased.[2] These fears carried through into the twentieth century, with the London coffin manufacturer Dottridge Bros continuing to market their "Life-Saving coffin" until around 1914.[3]

Those with the greatest fear of being buried alive used their will to stipulate the means by which physicians should check for signs of life. For instance, Jeanette Caroline Pickersgill, the first person to be cremated at Woking Crematorium in March 1885, stated in her will:

> I direct my executors after they have obtained a certificate (medical) of my death and before the coffin is closed to cause the arteries or large veins to be opened and I bequeath to my executors the sum of five pounds five shillings to be paid by them to Dr Langdon Down or any surgeon who may open the veins of my neck aforesaid.[4]

In this case, Dr. Down did not receive payment as Mrs. Pickersgill was subjected to an autopsy before cremation. Another example from 1909 also involved a physical assault on the corpse:

> In the will of Dame Katherine Millicent Palmer, of Dorney House, Buckinghamshire, who died on January 10th, aged sixty-six years, widow of Sir Charles James Palmer, leaving estate of the gross value of £2,876, with net personalty £2,598, the following direction is set down concerning her remains: 'I have a great horror of being buried alive; therefore I wish my finger to be cut, and bequeath £10 to Dr Wilmot or any other doctor who is attending me at my death for such service'.[5]

The nineteenth-century fear that death-timings could be mistaken can be traced to a deficiency in the Registration of Births and Deaths Act 1836.

This Act required a physician to supply a certificate confirming death, but it did not stipulate that a physical examination of the body must take place prior to its issue. On the medical certificate of the cause of death, the doctor could simply append "I am informed", leaving any person without medical qualifications or experience to state that life was extinct. As with all medical services at the time, the mandatory requirement for an examination would have involved the payment of a fee; for the poor, this would have represented a tax on death. Furthermore, tests for death would have been rudimentary and, perhaps, not necessarily conclusive. If a doctor was summoned, the diagnostic equipment in his bag would have comprised of little more than a stethoscope, a thermometer, an ophthalmoscope (for detecting decomposition in the retina), a hypodermic syringe (for injecting ammonia to detect inflammation), and a magnesium lamp (for examining circulation between the skin of the fingers).[6]

Among the poor, one fail-safe indicator that death had occurred was decomposition, especially the appearance of a green patch on the abdomen.[7] Keeping the deceased at home in the interval between death and burial would have provided sufficient time for this change to become apparent.[8] During this period, the undertaker would call to take a measurement and return to encoffin the body; a return visit may also have been necessary to seal the coffin if significant deterioration had taken place. It was not until the 1920s, that bodies were brought to the undertaker for more extensive treatment and transition to the chapel of rest, so an extended access to the body was important for assuaging fears of premature burial among the nineteenth-century poor.[9] Long periods between death and burial were not uncommon. In his account of working as a gravedigger in the Suffolk village of Akenfield from the interwar years onwards, William Russ indicated that bodies would be kept at home for up to twelve days, not only because people "didn't care to part with it," but also because they "were afraid the corpse might still be alive."[10]

THE CHANGING CONTEXT OF DISPOSAL

Population expansion in the nineteenth century presented numerous social and economic challenges. As regards the disposal of the dead, the 1830s saw the introduction of death registration and the end of the Church of England's near-monopoly on burial provision by the establishment of proprietary cemetery companies. By the 1850s, Burial Board cemeteries funded by local ratepayers were transforming the landscape of burial in urban areas. With the increase in places approved for burial, and an increase in the means for disposing of the dead, the interval between death and burial was shortened. This allowed fears of premature burial to flourish, causing further changes in services for the disposal of bodies. The next section considers five of the main changes that occurred in the late nineteenth century.

The first development was cremation. In January 1874, the publication of Sir Henry Thompson's seminal paper "The Treatment of the Body after Death" led to the founding of the Cremation Society of England (CSE) and the building, five years later, of the first crematorium at Woking (however, because of issues concerning the legality of cremation, it was not until March 1885 when the cremator was used for the first time).[11] At first, this alternative to burial was not popular; there were two further cremations in 1885, ten in 1886, and thirteen in 1887. By 1900, only 444 cremations took place at the four crematoria then in operation, representing only 0.07% of deaths in the UK.

The CSE were aware that cremation could be used to conceal crime so Thompson joined his fellow physicians in the Society (including the surgeon Sir Thomas Spencer Wells and Ernest Hart, the editor of the *British Medical Journal*) in devising the certification required to ensure the cause of death had been ascertained. The CSE was influenced by the French system in which a *Medecin Verificateur* was engaged by the state to confirm death; indeed, Sir Henry included an example of a form used in the French system in his 1899 book *Modern Cremation: Its History and Practice*. Based on this system, the Society required the completion of a series of certificates by three physicians: one to give the cause of the death; a second to confirm this information; and a third, appointed by the crematorium to be the "medical referee," who independently reviewed all the documentation. Initially, Thompson himself vetted all the documents in his capacity as the first Medical Referee at Woking. A slightly modified system of this death certification was adopted by all the other crematoria opening after Woking, and the usage of these documents became formalised in the Cremation Act 1902. Between 1885 and the commencement of the legislation in 1903, nearly 3,300 cremations took place using this documentation.

The second key development was the building of mortuaries. From the 1870s onwards, the medical press published numerous accounts of the insanitary conditions in which bodies were retained at home, along with the continued prevarication by parishes, districts, the Metropolitan Board of Works and the Home Office over the provision of mortuaries.[12] In 1875, the *British Medical Journal* surveyed facilities in London and found that out of the 20 districts that replied, just under half had no mortuary accommodation, whilst 13 had no post-mortem room.[13] The few mortuaries that had been opened, such as at Clerkenwell, Marylebone, and Bow, were poorly appointed and this put off the people who needed them most. A report on the health of Marylebone published in 1875, suggested that the use of the parish mortuary for storing coffins had declined. *The Lancet* remarked: "The duty of educating the poor to overcome their prejudice against using mortuaries is as clear as is the duty of the urban sanitary authorities to provide them".[14] This view was still prevalent in the 1920s when Bertram Puckle noted,

> The thought that the bodies of friends and relations should be taken to a mortuary suggests to the average mind an indignity, a social degradation. The mortuary is regarded as especially provided by the State for the bodies of unfortunate outcasts picked up from the gutter, or dragged from the river, or at the best, as a place where the suicide or a person meeting with some dreadful accident is impounded till a jury can be called together for an inquest. We associate it mentally with the prison and the workhouse.[15]

The Public Health (London) Act 1891 finally made it mandatory for every sanitary authority in the capital to provide a mortuary.[16] Mortuaries could take many different forms: in 1904, the Borough of Kensington constructed a "chapel of rest", as it was termed, in Avondale Park, Notting Hill.[17] Despite being located in an area of dense housing, its use was only modest, as was reported over 20 years later:

> We fear, however, that there is a still a tendency of the part of persons living in tenements of one, two or three rooms to retain in their tenements the bodies of deceased relatives awaiting burials, and that the accommodation afforded by the chapel was greatly overlooked.[18]

Despite being designated a "chapel of rest", this facility was on a similar footing to any other mortuary, being no more than a communal storage space provided by the local authority. It was, however, undertakers who addressed popular prejudice against mortuaries by opening private chapels of rest where the coffin could rest until the funeral. Furnished in an ecclesiastical manner and open to access without charge, such chapels can be found from around 1914, particularly in the urban areas. These early mortuaries and chapels of rest were, however, spaces to accommodate the dead; they were not akin to the "waiting mortuaries" provided in Paris and elsewhere in Europe, to which the dead were transferred until decomposition confirmed death.

The third point concerns change within the funeral industry. The two key functions of the nineteenth-century undertaker were the provision of the coffin, and the subsequent arranging of transportation to the place of disposal. Other goods and services were also supplied for those who could afford it, such as mourning wear (including hatbands and gloves). As already identified, the undertaker's contact with the dead body was no more than lifting it into a coffin, although occasionally embalmments took place for overseas transportation of the dead. In 1900, there was a turning point in the function of the undertaker when two embalming tutors toured Britain, leading sessions in the craft of arterial preservation of the dead, a technique already well-developed in the United States.[19] Practitioners of this method formed an association, the British Embalmers' Society, with the support of fluid manufacturers, from whom it was hoped undertakers would make purchases. *The Undertakers' Journal*, which had been founded in 1885 and regularly published articles penned by the leading US practitioners, lent its support to embalmers. However, as with cremation, change was slow to come to the funeral industry. Very little embalming took place in the early part of the twentieth century, and when it did, treatment was at home and was costly. Nevertheless, embalming gave undertakers a professional credibility by their acquisition of anatomical and sanitary knowledge, the founding of a qualifying association and, in 1908, the publication of a code of ethics. This trend toward professionalisation was symbolised by the creation of the British Undertakers' Association, with the primary objective being state registration.

The fourth aspect of changing services in the late nineteenth century concerns legislation relating to aspects of disposal. Whilst burial grounds became increasingly regulated, doctors were still not required to examine the deceased before giving a death certificate. The CSE promoted the Disposal of the Dead (Regulations) Bill 1884, which was designed to regulate cremation and also introduce a physical examination of the deceased, but their initiative failed. Meanwhile, in 1893, the Departmental Committee on Death Certification drew attention to the absence of mandatory inspection of the body and noted that "in some instance a skilled observer would only be able to pronounce whether life was extinct".[20] Again, few of its recommendations were implemented.

The fifth area was the reform of the funeral sector. Around the time that cremation was being promoted in the mid-1870s, the surgeon and etcher Sir Francis Seymour Haden defended the practice of burial against the cremationists. Haden argued that, if carried out correctly, using wicker or papier mâché "Earth to Earth" coffins interred in sandy, porous soil, would allow the body to deteriorate swiftly to its constituent elements. This would also allow others to reuse the grave.[21] Another attack on emboldened undertakers came from the Church of England Funeral and Mourning Reform Association, founded in 1875 by the Reverend Frederick Lawrence. This group concerned itself with the "excessive cost and cynical manipulation of funerals by undertakers".[22]

Within the funeral industry, the Paddington-based funeral director and proprietor of *The Undertakers' Journal*, Halford Lupton Mills, also pursued a reform agenda.[23] He encouraged the use of the open-sided horse-drawn hearse, rather than an elaborate closed carriage, and dismissed the sale of unnecessary paraphernalia, including the use of mutes and the carrying of trays of feathers. The trend for advertising a scale of charges for funerals in newspapers and trade directories also suggests that some undertakers were keen to distance themselves from the unscrupulous behaviour of some colleagues. Whilst the overall purpose of these reforms was to reduce funerary expenditure through simplified disposal practices, they did not, however, address the issue of certification, nor did they provide reassurance to those concerned by the possibility of being buried alive. It was the latter that led to the founding of a pressure group specifically for this purpose.

Premature Burial and the Undertakers

In 1896, the London Association for the Prevention of Premature Burial (LAPPB) was established by William Tebb and a Gloucester-based general practitioner, Dr. Walter Hadwen. Tebb and Hadwen were stimulated by accounts of vivisepulchre in the popular press during 1895, and the following year Tebb wrote a book on the subject with Edward Vollum entitled *Premature Burial and How It May Be Prevented*. Both Tebb and Hadwen were on the fringes of medical orthodoxy because of their support for anti-vaccination, anti-vivisection, and vegetarian causes. Between

1901 and 1914, the Association intervened in debates on funerary practices through its journal, the *Burial Reformer* (renamed *Perils of Premature Burial* in 1909).

Despite the generous coverage of LAPPB in *The Undertakers' Journal*, it mirrored the funeral sector as a whole in oscillating between expressions of ridicule and support for the movement. For instance, in 1898, the editor commented that "Premature burial, [was] an unwholesome and unnecessary subject in England",[24] whilst previously stating that "Tenders for the supply of premature burial stories are not invited at this office".[25] In 1909, when the title of the *Burial Reformer* changed, the editor of *The Undertakers' Journal* suggested as alternatives *The Medical Sinecurist* (insinuating that physicians could earn a healthy income from certifying deaths) or *The Burial Fiction Monthly*.[26] In its journal, the LAPPB published many accounts of people who were buried alive.[27] Of these, only a case of catalepsy in 1905 contained any measure of credibility. This involved a young woman who was certified dead by a physician acting on information provided by the family. Luckily, she was discovered alive by an undertaker taking a measurement for her coffin.[28] The response of *The Undertakers' Journal* was to publish letters requesting definite cases of premature burial. The LAPPB must have been very disappointed when Dr. Frederick Waldo, the South London Coroner, stated that he had come across no proven cases of premature burial.[29]

In the year following the founding of the LAPPB, the Secretary of State for the Home Office was asked about a *Lancet* report that stated that 15,000 people were buried annually without medical certificates.[30] This figure was denied and it was further stated that no cases of premature burial had been brought to the attention of the Home Office. This did not persuade Sir Henry Thompson who pointed to a national problem:

> Previous to cremation, let me say that it is *sine qua non* that a careful examination of the body by two medical practitioners (neither of whom is related to the deceased) must be made, and the cause of death clearly stated...In England and Wales an average of fifteen thousand are buried annually without a certificate of any sort, and the proportion is much larger in Scotland, amounting to about 50 per cent.[31]

The lack of mandatory death certification was an issue that energised undertakers, and those concerned about premature burial. In 1905, the British Undertakers' Association (BUA) was founded, and it joined the LAPPB in addressing the issue. However, for the undertakers, this campaign was part of their desire to offer more services and develop professional credibility; the LAPPB preferred to raise awareness of the deficiencies in the certification system through sensational reporting. Despite this, *The Undertakers' Journal* backed up some of the claims of the LAPPB in 1908, when it was stated that thirteen cases of premature burial were reported in the previous year.[32]

Despite this sensationalism, practical proposals did emerge. Inspired by examples in France and Germany, the LAPPB campaigned for the building of "waiting mortuaries" where the corpse could rest until signs of decomposition proved death had occurred. With grudging use of public mortuaries as they were, there was certainly no appetite for this initiative. The following year, the LAPPB suggested that the occupier of a household should be obliged to instruct a Medical Officer of Health to remove a body into their care prior to burial or cremation.[33] Encouraged by a proposed bill in Massachusetts in 1903, they drafted legislation that floated the idea of a "death verifier" whose role would be to allow burial only after the body showed signs of decomposition.[34]

On the part of the undertakers, the funeral director James Broome argued that medical examiners [death certifiers] were too expensive and that people were not insured if they used their services. He proposed a solution: "Where uncertainty of death exists...call in a member of the British Embalmers' Society who is conversant with signs and tests".[35] Broome recommended that all funeral directors carry out tests for death, while the founding father of the British Undertakers' Association and embalming pioneer, Henry Sherry, appropriately suggested that embalmers should be trained in resuscitation.[36] In 1909, a "Death Registration and Burial Bill" was discussed at the BUA convention by Albert Cottridge, a London funeral director and advocate of embalming and the registration of funeral directors. Cottridge argued that each district should appoint a public certifier of deaths; that no death should be registered unless a medical certificate had been completed following examination of the deceased; that a physician completing a medical certificate should be paid 2s 6d (12½p); and that the person registering the death should receive a certificate from the registrar, then hand this to the cemetery for endorsement before being returned to the registrar. The registration of stillbirths was also to be included.[37]

Outside the funeral sector, weaknesses in the certification system were highlighted. In 1910 a *John Bull* article by the barrister and legal expert on burials, Alfred Fellows, raised the issue, while the following year the Coroners' Law and Death Certification (Amendment) Bill sought to introduce recommendations from the 1893 and 1908 Select committee on death certification and Coroner's legislation respectively. None of these efforts, however, yielded any major change and there appeared to be little in the way of parliamentary or medical support. In 1910, *The Undertakers' Journal* wearily noted that "The present law of death certification offers every opportunity for premature burial and every facility for concealment of crime".[38]

The outbreak of World War I curtailed the enthusiasm of the LAPPB, but by 1919 it was still calling for a public certifier of deaths. One of its members, the Reverend Hugh Chapman, chaplain of the Savoy Chapel, suggested that if a Cabinet Minister were to be buried alive, public attention might be awakened to the issue.[39] In 1922, a less sensational comment came from the Association's president, Sir George Greenwood MP, who declared that the present certification system was "a disgrace and a national danger". The following year, it was the turn, once again, of the undertakers to propose legislation, and the BUA's secretary, James Hurry, suggested that all undertakers enlist the support of their MPs to change the law. He proposed:

> That it shall be compulsory for medical men to view the body after death before granting a certificate
>
> That the cause of death shall be placed on the certificate by the Registrar
>
> Registration of still-born children.[40]

This coincided with the undertakers also drafting legislation for state registration containing a clause that members must attend a course of study in sanitation (embalming) and care of the deceased.[41]

Further proposed bills to improve the certification system were discussed by the BUA in 1922, whilst the President of the British Medical Association, Professor David Drummond, advocated a compulsory post-mortem examination. This was a certain, but costly and controversial way of preventing premature burial.[42] Some improvement came with the Births and Deaths Registration Act 1926, particularly in respect of stillbirths and certification before disposal, but there was no broad political support for the public certifier and still no requirement for a doctor to see the body after death, despite the insertion of a clause requiring this during the third reading of the bill.[43]

By the interwar years, the landscape of funerals and the work of the funeral director had changed. In 1927, a second association, the British Institute of Embalmers (BIE), was founded to provide formal training and a qualification. As the number of practitioners and availability of the treatment increased, the cost decreased and more funeral directors advocated embalming. Recognising the issue of premature burial, the BIE syllabus commenced with instructions on how to test for death before treatment. Whilst the comment of the American embalmer, O.K. Buckhout, that "embalming prevents the possibility of premature burial" was correct, the consequence of raising and injecting the carotid artery if the person was still alive would have been catastrophic.[44] Testing for death by funeral directors and embalmers was logical as they had the experience of handling the dead on a routine basis. Cottridge included a chapter on the subject in his book *Anatomy and Sanitation*, published in 1925, while Medical Officers of Health addressing the annual BUA convention frequently referred to the need for the tests to be carried out.[45] The merits of the different tests often became a discussion point within the pages of the trade journals.[46]

In this changed landscape, testing for death was a service that gave people confidence that funeral directors and embalmers could treat the body appropriately. This confidence was especially important in the 1930s when it was claimed that 60% of deaths were certified without the doctor having seen the body.[47] Despite some support from general practitioners for their practice, funeral directors and embalmers did not receive universal backing from the British medical community. Not only did training for embalming lack external accreditation, but embalming tutors did not typically have medical experience. But perhaps the key reason for the reluctance of the medical community to approve of funeral directors having a role, similar to the French *Medecin Verificateur* was that this would have usurped the doctor's presence at the dying person's bedside.

Despite the lack of formal validation in their quest, undertakers continued to be interested in the tests for death. In 1927, a dye was demonstrated at the BUA Convention that could be injected into the body to determine whether life was extinct. The effects of the "infallible and harmless" *Obiturin* were discussed in a report of the proceedings:

> Previous to the lecture Mr AG Hurry had submitted himself to an experimental demonstration to the effect of *Obiturin* on the living human body. The injection was made in Mr Hurry's forearm while a like experiment was performed upon the lecturer by Ald [Alderman] Kenyon. The result was awaited with great interest and those present had the satisfaction of observing an interesting reaction, Mr Hurry going round the room in order that all present might examine at close quarters the green discolouration which proved that he was very much alive.[48]

By the late 1930s, physicians were required to examine bodies and sign death certificates in the increasing number of state institutions in Britain. This trend continued particularly after the founding of the National Health Service in the late 1940s. Despite a continued low preference for cremation (in 1927 only 0.59% of deaths were followed by cremation), people could now be confident that premature burial was unlikely given the requirement for two physicians to carry out a careful external examination of the body.[49] However, while there is evidence to show that in urban areas the preference for cremation shifted dramatically when local facilities were provided, it would be another 40 years before the trend spread outside the cities. It was only in 1965 that cremations outnumbered burials for the first time.[50]

The LAPPB, meanwhile, limped along into the 1930s.[51] Hadwen died in 1932, and three years later the Association become affiliated with the National Council for the Disposition of the Dead (NCDD), an organisation with an agenda to promote cremation along with the registration of funeral directors. However, the minutes indicate that by the second meeting, the LAPPB was not listed as a participant, and the NCDD failed in any case due to lack of support.[52] The quest for a public certifier and the construction of "waiting" mortuaries was never realised, particularly as by the 1940s the vast majority of funeral directors had opened chapels of rest in their private premises, where more embalming was taking place. Remarkably, the law was never changed to require a doctor to examine a person after death, unless cremation was called for.

While concerns about premature burial in the nineteenth century had their basis in a deficiency in legislation and medical practice, for most people, declaration of the fact of death would often have come from informal care-givers who nursed the dying. Their ability to recognise the boundary between life and death was solely dependent on past experience. Given the landscape of death and burial in Britain today, with professional funeral directors managing the treatment and disposal of most corpses, the informal knowledge about bodies and decomposition that people, especially the poor, possessed is worth reflecting on. Yet, the number of cases where the signs of death were misinterpreted needs to be viewed in perspective. While the LAPPB highlighted only a few factually proven incidents, the scale of the issue was unknown. It was one thing if a person was found to be living before burial, but quite another if a person was found to have lived after interment had taken place. Theoretically speaking, without mass exhumations, it is impossible to know how many bodies show signs of life after interment. Undertakers seized on this macabre situation as an opportunity to exercise power over the body by becoming the self-appointed person to carry out the tests for death. Although it was not easily gained, this power gave them a new status as quasi-medical practitioners and helped shed the Dickensian image of disreputability inherited from their nineteenth-century forebears.

Today, the possibility of premature burial in Britain has been diminished by several factors: Due to shifts in the culture of death and dying, a large proportion of deaths now take place in institutions where physicians with a range of diagnostic equipment are on hand to confirm death. Furthermore, the majority of deaths are now followed by cremation, and this requires two doctors to examine the deceased. In addition, the interval of 5–14 days between a person's death and funeral, along with the widespread adoption of embalming, has reduced the possibility of a premature burial. Nevertheless, knowledge of the tests for death still remains part of the Diploma in Funeral Directing examination in Britain, while the National Association of Funeral Directors' *Manual of Funeral Directing* states: "Every funeral director should be able to satisfy himself, and on occasion the family, that death has actually taken place".[53] However, as legislation still does not require a doctor to examine the deceased before completing the Medical Certificate of the Cause of Death, the issue continues to engage the minds of funeral directors, if not the general public.

Notes

1. George K. Behlmer, "Grave Doubts: Victorian Medicine, Moral Panic, and the Signs of Death", *Journal of British Studies*, 42:2 (2003), p. 207. See also Jan Bondeson, *Buried Alive: The Terrifying History of Our Most Primal Fear* (New York, 2001); Joanna Bourke, *Fear: A Cultural History* (London, 2005).
2. Ralph H. Major ed., "Devices to Prevent Premature Burial", *Journal of the History of Medicine and Allied Sciences*, 3:1 (1948), pp. 161–171.
3. "The Life-Saving Coffin", *The Undertakers' Journal* [hereafter *TUJ*], 36:2 (1912), p. 38. See also "Coffin for the Living", *TUJ*, 20:5 (1905), p. 116.
4. Will dated 20 December 1882, copy in author's collection.
5. "Horror of Premature Burial", *TUJ*, 24:8 (1909), p. 52. See also "Making Sure of Death", *TUJ*, 24:5 (1909), p. 105; "Fear of Premature Burial", *TUJ*, 42:9 (1927), p. 318.
6. Behlmer, "Grave Doubts", p. 217.
7. "Death Certificates: Urgent Need of Reform of the Law", *TUJ*, 26:11 (1911), p. 296. See also "Fear of Premature Burial", *British Undertakers' Association Monthly* [hereafter *BUAM*], 10:10 (1931), p. 232.
8. See Julie-Marie Strange, *Death, Grief and Poverty in Britain, 1870–1914* (Cambridge, 2005).
9. See Brian Parsons, *The Undertaker at Work: 1900–1950* (London, 2014).
10. Ronald Blythe, *Akenfield: Portrait of an English Village* (New York, 1969), p. 313.
11. See Brian Parsons, *Committed to the Cleansing Flame: The Development of Cremation in Nineteenth-Century England* (Reading, 2005).
12. See Pam Fisher, "Houses for the Dead: The Provision of Mortuaries in London, 1866–1889", *The London Journal*, 34:1 (2009), pp. 1–15.
13. See "Mortuaries for the Metropolis," *The British Medical Journal* (1875), pp. 802–803.
14. "Prejudice against Mortuaries", *The Lancet* (1875), p. 331.
15. Bertram S. Puckle, *Funeral Customs: Their Origin and Development* (London, 1926), p. 28.
16. Albert Chambers Freeman, *The Planning of Poor Law Buildings and Mortuaries* (London, 1906); "A Public Mortuary and Post-Mortem Room", *The Lancet* (1870), p. 318.
17. Borough of Kensington Council Minutes, 29 March 1904, p. 236; 5 July 1904, p. 346. The Royal Borough of Kensington & Chelsea, Local Studies Library.
18. Ibid., 18 January 1927, p. 111.
19. See Parsons, *The Undertaker*.

20. *Reports from the Select Committee on Death Certification* (London, 1893), p. xiii.
21. Parsons, *Committed to the Cleansing Flame*, pp. 204–220.
22. Ibid., p. 212.
23. See Brian Parsons, "Halford Mills: Funeral Reformer and Pioneer of Embalming", *Funeral Service Journal* (2005), pp. 64–72.
24. "Notes", *TUJ*, 13:6 (1898), p. 83.
25. "Notes", *TUJ*, 11:6 (1896), p. 82. See also "Editorial: How to Prevent Premature Burial", *TUJ*, 18:7 (1903), pp. 165–166.
26. "Notes", *TUJ*, 24:3 (1909), p. 50.
27. "Burial Alive", *TUJ*, 13:8 (1898), p. 124; "Notes", *TUJ*, 13:11 (1898), p. 159; "Reported Cases of Premature Burial", *TUJ*, 18:5 (1903), p. 110; "Recent Live Burial Cases", *TUJ*, 19:5 (1904), p. 97.
28. See "Notes", *TUJ*, 20:2 (1905), p. 26; "The Accrington Catalepsy Case", *TUJ*, 20:2 (1905), p. xi.
29. "Letter to the Editor. Burial Alive", *TUJ*, 16:8 (1901), pp. 188–189; "Notes", *TUJ*, 19:2 (1904), p. 36.
30. *The Lancet* (1897), pp. 1155–1156.
31. "Cremation", *TUJ*, 12:7 (1897), p. 95.
32. "Editorial. Premature Burial", *TUJ*, 23:2 (1908), pp. 35–36; "Premature Burial. Thirteen Cases During the Past Year", *TUJ*, 23:2 (1908), p. 36; "Notes", *TUJ*, 23:3 (1908), pp. 53–54.
33. "A Bill to Prevent Alive Burial", *TUJ*, 16:6 (1901), pp. 136–137.
34. "Notes", *TUJ*, 18:12 (1903), p. 260; "Notes", *TUJ*, 19:12 (1904), p. 245; "Notes", *TUJ*, 20:1 (1905), p. 2.
35. "Letter to the Editor. The Prevention of Premature Burial", *TUJ*, 20:1 (1905), p. 16.
36. H. Sherry, "The Sanitary Advantages of Modern Embalming", *TUJ*, 20:8 (1905), pp. 188–189.
37. A.J.E. Cottridge, "The Death Registration and Coroner's Bill", *TUJ*, 24:7 (1909), pp. 171–172.
38. A. Fellows, "Death Certificates: The Weaknesses of the Present Law", *TUJ*, 25:11 (1910), p. 266; "Coroners' Law and Death Certification (Amendment) Bill", *TUJ*, 25:12 (1910), pp. 292–294.
39. "Premature Burial", *BUAM*, 1:11 (1922), p. 251.
40. "Registration of Deaths", *BUAM*, 2:9 (1923), p. 248.
41. "Registration", *BUAM*, 1:11 (1922), pp. 250–251.
42. "Post-Mortem Examinations for All", *BUAM*, 1:4 (1922), p. 74.
43. "Births and Deaths Registration Bill", *TUJ*, 41:3 (1926), p. 73; "Notes", *TUJ*, 41:4 (1926), p. 105. See also "Births and Deaths Registration Bill", *Hansard*, 26 February 1926, Cols 977-998; "Births and Deaths

Registration Bill' *Hansard*, 18 June 1926, Cols 2684-2692; S.A. Smith, "Cremation and Crime", *TUJ*, 45:9 (1930), pp. 295–298.
44. "The Art and Use of Embalming", *TUJ*, 15:7 (1900), p. 78.
45. A.J.E. Cottridge, *Anatomy and Sanitation* (London, 1925), pp. 103–104; "Proposed Syllabus of Instruction in Sanitation", *BUAM*, 2:1 (1922), p. 2.
46. E.P. Vollum, "Last Tests for Death", *TUJ*, 19:4 (1904), pp. 80–81. See also F.A. Sharpe, "Modes of Death", *BUAM*, 1:12 (1922), p. 276; A.J.E. Cottridge, "Signs of Death", *BUAM*, 11:4 (1931), pp. 82–83; G.C.F. Rose, "Tests for Cause of Death", *BUAM*, 11:8 (1932), p. 183.
47. "Editorial: Imagination and Progress", *TUJ*, 45:2 (1930), p. 56.
48. "Lecture on Obiturin", *BUAM*, 7:2 (1927), pp. 44–46.
49. See also "Premature Burial", *The Times*, 31 March 1927; "The Federation of Cremation Authorities in Great Britain", *TUJ*, 42:7 (1927), pp. 245–246.
50. See Peter C. Jupp, *From Dust to Ashes: Cremation and the British Way of Death* (Basingstoke, 2005).
51. "Personal", *The Times*, 10 July 1930; "News in Advertisements!", *The Times*, 24 June 1931.
52. Jupp, *From Dust to Ashes*, pp. 114–115; "Prevention of Premature Burial!", *The Times*, 5 February 1936; "National Council for the Disposition of the Dead", *BUAM*, 14:12 (1935), p. 246.
53. *Manual of Funeral Directing* (London, 1976), p. 45.

Author Biography

Brian Parsons has worked in funeral service in London since 1982. His Ph.D. focused on the impact of change, during the twentieth century, on the funeral industry. He is the author of *The London Way of Death*, *Committed to the Cleansing Flame: The Development of Cremation in Nineteenth Century England* and *The Undertaker at Work: 1900–1950*.

Open Access This chapter is licensed under the terms of the Creative Commons Attribution 4.0 International License (http://creativecommons.org/licenses/by/4.0/), which permits use, sharing, adaptation, distribution and reproduction in any medium or format, as long as you give appropriate credit to the original author(s) and the source, provide a link to the Creative Commons license and indicate if changes were made.

The images or other third party material in this chapter are included in the chapter's Creative Commons license, unless indicated otherwise in a credit line to the material. If material is not included in the chapter's Creative Commons license and your intended use is not permitted by statutory regulation or exceeds the permitted use, you will need to obtain permission directly from the copyright holder.

CHAPTER 6

The Death of Nazism? Investigating Hitler's Remains and Survival Rumours in Post-War Germany

Caroline Sharples

"Becoming really dead", argues Thomas Laqueur, "takes time".[1] It has been more than 70 years since Adolf Hitler's suicide in his Berlin bunker, yet the passage of time has done little to diminish public fascination with the Nazi leader, nor stem speculation surrounding the circumstances of his demise. Indeed, some people have doubted whether Hitler died in Berlin at all; survival myths remain popular fodder for tabloid newspaper articles, sensationalist television documentaries, and best-selling books.[2] Fundamentally, the endurance of such legends is rooted in the chaos of the immediate post-war era and the Allies' failure to positively identify any human remains as those of the former *Führer*. In the absence of a body, what counts as irrefutable proof of death?

In 1945, the western Allies' answer was to establish a clear timeline of the events leading up to the suicide, piecing together witness testimonies from Hitler's staff and poring over key documents, such as his last will and testament. The first history on this topic, produced by

C. Sharples (✉)
University of Roehampton, London, UK
e-mail: caroline.sharples@roehampton.ac.uk

Hugh Trevor-Roper in 1947, reflected this approach, using information collected during the author's service with British Military Intelligence.[3] Potential forensic evidence, gathered by the Soviets, was released only gradually. It was not until 1968 that Lev Bezymenski was able to publish his account based upon the autopsy reports on the alleged remains of Hitler and Eva Braun.[4] Since the end of the Cold War, additional material from the former Soviet archives has revived scholarly interest in the case, spurring reassessments of the available medical evidence by the likes of Ada Petrova, Peter Watson, and Daniela Marchetti.[5] Yet, while there are now detailed—if varying—accounts of the *mode* of Hitler's demise, there has been little attempt to explain the origins and persistence of survival myths, or to locate Hitler's end within the broader context of a National Socialist fixation with the dead.

This chapter, therefore, sets out to demonstrate that the death of Adolf Hitler was both a biological and social process. The Nazi regime had been constructed around a cult of personality and the leader's death became synonymous with Germany's total defeat in the Second World War, a significant rupture marking the end of National Socialism itself. In reality, of course, the regime limped on for an additional eight days without Hitler, and supposed sightings of the former leader kept his memory very much alive in the public imagination. Hitler's suicide, then, was hardly a "zero hour" for the nation, but an event that serves to demonstrate the complexity of post-conflict commemorative culture.[6] Drawing upon British Foreign Office and Military Intelligence records, this chapter traces the Allies' efforts to sort the fact from fiction. At the same time, it also reveals how post-war power struggles to control the narrative of Hitler's death contributed to the subsequent survival mythology, with Nazis and Allies both deliberately casting doubt on the timing and cause of death to further their own interests.

To understand initial German reactions to the loss of Hitler, we have to situate them within a longer history of Nazi rituals and martyrdom legends.[7] During the Third Reich, the Nazi regime routinely peddled the notion that fallen comrades were not truly dead, but continued to fight for Germany as part of an immortal, spiritual army. This was important, ideological glue for manufacturing the *Volksgemeinschaft* (People's Community) and preparing the population for the necessary challenges ahead. The anniversary of the 1923 Munich Putsch, in which 16 Nazis had been killed, became one of the holiest days in the Nazi calendar. Speaking at the commemorations in 1942, for example, Hitler declared:

Truly these sixteen who fell have celebrated a resurrection unique in world history... From their sacrifice came Germany's unity, the victory of a movement, of an idea and the devotion of the entire people...All the subsequent blood sacrifices were inspired by the sacrifice of these first men. Therefore we raise them out of the darkness of forgetfulness and make them the centre of attention of the German people forever. *For us they are not dead.* This temple is no crypt but an eternal watch. Here they stand for Germany, on guard for our people. Here they lie as true martyrs of our movement.[8]

This existing emphasis on the eternal spirit of Nazism constituted a ready-made framework for casting doubt on Hitler's own mortality. In addition, the German public had become somewhat accustomed to Hitler being able to extricate himself from perilous situations. Hitler had survived numerous assassination attempts during his time in power—most notably, Georg Elser's bombing of the Bürgerbräukeller in Munich in November 1939 and the Operation Valkyrie attempt in the Wolf's Lair in July 1944. Following the latter event, Hitler gave a radio speech in which he declared that his survival was proof that his work was blessed by Divine Providence.[9] Given this background, it is understandable that his eventual, ignoble end in a Berlin bunker may have been viewed with disbelief.

One of the key challenges facing the Allies in 1945, then, was to dismantle some of these prevailing mythologies. A thorough denazification programme was intended to cleanse Germany of every last vestige of National Socialism, including the removal of Nazi symbols from the landscape. The elaborate memorials that had been constructed in honour of the "old fighters" killed in Munich were removed and the iron sarcophagi that had housed their mortal remains were recycled for use in repairing regional railway lines. "Ordinary" cemeteries were also affected by the political transition away from fascism: gravestones were purged of swastikas and other Nazi imagery or, in some cases, destroyed altogether. The Allies' central aim was to prevent the formation of pilgrimage sites that could be used to sustain National Socialist ideology. Consequently, those who had died fighting for Nazism were now being subjected to a form of "social death", stripped of their previously exalted status with their past achievements now rendered taboo in public discourse.[10] The fate of Hitler himself quickly became entangled with this denazification process. With his image banned after the war, and access to the former Reich Chancellery and bunker controlled by the Allies, the German people had little outlet for mourning their fallen leader. This may have come as something as a culture shock after the sophisticated state funerals of the Third Reich. Unlike the posthumous history of other dictators, such as Stalin or Mussolini, there was no public memorial or display of Hitler's body. Consequently, John Borneman argues that the population endured "an enforced silence about the scene of death and the whereabouts of the corpse".[11] The extent of this "silence" can, of course, be called into question by the sheer number of rumours that emerged immediately over the timing, manner, or actuality, of Hitler's death.

It was at 10.30pm on Tuesday, 1 May 1945, following three solemn drum rolls, that Grand Admiral Karl Dönitz took to the airwaves of North German radio to make a crucial announcement: "German men and women, soldiers of the armed forces: our Führer, Adolf Hitler, has fallen. In the deepest sorrow and respect, the German people bow".[12] Reflecting on the manner of Hitler's death, Dönitz added:

> At an early date, he had recognised the frightful danger of Bolshevism and dedicated his existence to this struggle. At the end of his struggle, of his unswerving straight road of life, stands his hero's death in the capital of the German Reich. His life has been one single service for Germany.[13]

Further reports within the German press the following day elaborated on the glorious nature of the Führer's last stand—and applied a similar rhetoric of immortality to that previously assigned to those killed in the Munich Putsch. The *Hamburger Zeitung*, for example, insisted:

> We know that he must have perished while fighting bitterly in the Reich Chancellery. We know that the enemy will be able to find a body in the ruins caused by countless artillery shells and countless flame throwers, and *that they may say that it is the Führer's body, but we will not believe it*...What is mortal of him has perished, has passed away but he has fulfilled his most beautiful oath [to give his life to his people]...He began by fighting for his people, and he ended that way. A life of battle.[14]

Similarly, a message broadcast to troops stationed in the Netherlands proclaimed: Adolf Hitler, *you are not dead*, you live on within us. The ideals which you gave us cannot be extinguished ... Beneath the ruins of a devastated Berlin, you remain the fountain of all Germans.[15]

In terms of the final pieces of Nazi propaganda, then, the cult of the Führer remained very much alive. His memory and, in particular, the seemingly dramatic nature of his demise—courageously resisting the Soviet advance into Berlin—served as a last-ditch appeal to the German people to keep on fighting. These descriptions of Hitler's final moments, though, were designed to obscure the truth. The consensus of scholarly opinion and witness testimony suggests that, on 30 April 1945, Hitler chose to kill himself rather than end up in the hands of the advancing Russians. In his last hours, he married his long-term companion, Eva Braun, dictated his will and political testament, and administered cyanide to his beloved Alsatian dog, Blondi, to determine the effectiveness of the poison.[16] Having heard about the public desecration of Mussolini's corpse on 28 April, he made preparations to ensure that no similar humiliation would be extended to his remains. Petrol was ordered and his staff members were instructed to incinerate his body when the time came. Indeed, Hitler's own precautions would prompt much of the post-war debate and confusion about his fate.

Almost immediately, the veracity of Dönitz's account was called into doubt by the Allies, and even some high-ranking Nazis. The day after Dönitz's radio address, the Russian newspaper, *Pravda*, proclaimed the whole story to be a "fascist trick to cover Hitler's disappearance from the scene".[17] Observers in Britain and the United States, while noting that a death fighting against the "Bolshevik hordes" would have been "quite in character" for Hitler, quickly moved to undermine what was left of the German war effort by issuing statements challenging Dönitz's account of Hitler meeting a "hero's death" in Berlin.[18] To support their claims (and to try and avoid their comments being dismissed as enemy propaganda), the western Allies seized upon an account of Hitler's failing health promulgated by the head of the SS, Heinrich Himmler, more than a month earlier. According to notes of a conversation between Himmler and the Swedish diplomat Count Folke Bernadotte on 24 March 1945, Hitler was "finished". Himmler claimed that the Führer was suffering from a brain haemorrhage and would be dead in a couple of days, if he wasn't already—a sentiment that immediately cast doubt on the precise timing of Hitler's demise.[19]

For the Allies, disseminating Himmler's version of events could sow the seeds of discord among the remnants of the Nazi leadership and shatter any remaining illusions that the general population still harboured about their "courageous" leader, preventing the formation of martyrdom myths. A Foreign Office memorandum noted that "there is every indication that German propaganda will play up the manner of Hitler's death with a view to establishing the Hitler legend. We must do all in our power to play it down".[20] Himmler's account was privately regarded as a "good weapon" to encourage the Wehrmacht, now released from their oath of loyalty, to surrender and prompt the fall of more German cities. A public statement issued by General Eisenhower dismissed Dönitz's statement as an effort "to drive a wedge between the British and Americans on one side and the Russians on the other".[21]

For Himmler, meanwhile, the original assertion in March 1945 that Hitler was in no fit state to rule served to strengthen his own negotiating hand for surrender, enabling him to present himself as the provisional leader of the country. Himmler was conspicuously absent from the public discussion of Hitler's death on 1 and 2 May, suggesting the continuance of a power struggle between himself and Dönitz. By advancing competing accounts of Hitler's health, the pair cast doubts on the closeness of one another's relationship with the Führer, and their right to rule in his stead.[22] At the same time, with one eye undoubtedly on the future, even Dönitz was rather muted in his eulogy, dedicating just six sentences of his radio broadcast to dealing with Hitler's death. Having been named as Hitler's successor, Dönitz then used the remainder of his radio broadcast to try and rally popular support behind him. Observers within the British Foreign Office similarly noted an absence of "fanatical party statements" in remembrance of their leader. Given the dire military situation, this relatively restrained response from Hitler's fellow Nazis may be seen as an attempt to dissociate themselves from the failing regime, and an effort to strengthen their own position with the advancing Allies. Different parties, then, were able to appropriate Hitler's death to further their own political cause.

Publishing Himmler's comments in early May 1945 sparked a long-standing fascination with Hitler's medical history, including the lingering physical effects of the attempt on his life in July 1944, and the psychological strain of living in the Berlin bunker during the final phases of the war. In the immediate aftermath of the war, the Allies initiated a search for any surviving medical records, and interrogated anyone who had treated Hitler in the past, knowing that such evidence could play a vital role in identifying any human remains. How and where Hitler died consequently became the subject of great speculation: was it inside or outside of the Führer bunker? Was it the result of a stroke or nervous collapse, cyanide capsule, lethal injection or gun? Could Hitler have taken cyanide and still have time to shoot himself in the temple? Had death occurred at Hitler's own hand, or was it the result of his doctor's intervention? Timing too, became a crucial issue. British military intelligence took great pains to reconstruct Hitler's movements in the final days and hours leading up to his death. But had death occurred even earlier than 30 April 1945? In June, the Allies received what they acknowledged to be a "very odd" communication from an Austrian builder to the effect that Hitler had actually been shot by an army general in March 1944, that the infamous July bomb plot later that year had been contrived by Nazi propagandists and that his corpse actually lay in a secret crypt below Obersalzburg, Hitler's mountain retreat in Berchtesgaden.[23] American investigators in Bavaria, however, could find no evidence to support this claim.

Had Hitler died at all? Amidst the Dönitz-Himmler debate in early May 1945, the *Daily Telegraph* published the testimony of Major Erwin Giesing, Hitler's personal physician, who refuted claims that the Nazi leader had been in ill health. In conclusion, the newspaper declared there was "some doubt" about the cause of Hitler's death, adding, "if he *is* dead".[24] By 15 May 1945, Winston Churchill had similarly admitted to the House of Commons that he was unable to confirm "beyond doubt" whether Hitler was dead.[25] The Chief of the US Secret Service, Brian Conrad, conceded that "the only decisive evidence … would be the discovery and positive identification of the corpse". He added, "if such evidence is unavailable, all that remains are the detailed accounts of certain witnesses who either knew of his intentions or were eyewitnesses to his fate".[26]

In terms of the former, the Allies soon appeared to have found what they were looking for. On 2 May—one day after Dönitz's radio address—Soviet forces occupied the former Führer bunker in Berlin and quickly discovered the remains of Propaganda Minister Joseph Goebbels, his wife Magda, and their six children. At the time, two Soviet officers, Lozovski and Litvinov, expressed some scepticism about the chances of finding Hitler's body too, believing that he had "gone to earth" along with Göring and Himmler.[27] On 5 May, however, the badly-burned corpses of a man and a woman were found in a bomb crater within the garden of the former Reich Chancellery, prompting speculation that they were that of Adolf Hitler and Eva Braun. Subsequent examination by Soviet forensic pathologists confirmed the presence of glass splinters in their mouths, consistent with biting into a cyanide capsule.[28] The male corpse was "heavily charred" and missing part of its cranium, but estimated to be "somewhere between 50 and 60 years" old; Hitler turned 56 in 1945. The other key point of interest for investigators concerned the male corpse's teeth, described as having "much bridgework, artificial teeth, crowns and filings".[29] Hitler's former dentist, Dr Hugo Blaschke had already managed to flee Berlin but, under Soviet interrogation, two of his former staff members, Käthe Heusemann and Fritz Echtmann, were able to describe and sketch Hitler's distinctive dental work from memory. On 9 May, they were invited to examine the physical remains retrieved from the bomb crater and concluded that they did, indeed, belong to the Nazi leader. Accordingly, on 31 May, KGB officer Ivan Serov informed Stalin and Molotov that "there is no doubt that the supposed corpse of Hitler is really his".[30]

While the official Soviet records were not released at this time, news of the discovered corpses was relayed in the media.[31] In June 1945, *The Times* also published a detailed account by Hermann Karnau, a former guard, who confirmed that he had seen the bodies of Hitler and Eva Braun lying in the grounds of the Reich Chancellery: "both bodies were on fire, but were clearly recognisable".[32] Yet this was not to be the end of the matter as the Soviets spent the rest of the summer of 1945 suddenly casting doubt on their own findings. On 10 June, Marshal Zhukov of the Red Army told a press conference: "The situation is very mysterious ... We have failed to discover a body confirmed as Hitler's. I cannot say anything definite about Hitler's fate".[33]

Rumours now spread that the charred remains previously seized upon by investigators had belonged to a body double and that Hitler had managed to flee the ravaged capital after all. On 5 July, a *Daily Telegraph* correspondent visiting the scene agreed that the previous narrative of suicide and cremation seemed doubtful:

> The account of Hitler's death in the shelter and the burning of the body, as told by the German policeman Kernau [sic] at 21st Army Group HQ recently, fits in perfectly with the evidence on view here. There are even five petrol cans, all marked with the SS sign...Corroboration is so overwhelming as to be almost suspicious.[34]

Why did the Soviets refute the dental evidence? The consensus among historians, including Russian scholars Vinogradov, Pogonyi and Teptzov, and the British academic Roger Moorhouse, is that this was a typical, cynical move by Stalin. In part, it reflected his own paranoia and mistrust of the forensic evidence being set before him; but it also became another way of exercising a degree of power over the other members of the wartime alliance. In July 1945, *The Times* repeated the claim that the jawbone found on the grounds of the Reich Chancellery had been positively identified as that of Adolf Hitler, but acknowledged that:

> Whatever pronouncement is made, it is certain that many people in Germany, especially here in Berlin, will go on believing in the legend of his escape under cover of one of the doubles he is supposed to have employed. It seems strange that of all the people of authority round Hitler, none has been found to give an account of what happened, and the circumstantial evidence accumulated from lesser fry could well be an attempt to cover Hitler's trail.[35]

The *Daily Herald* concurred, noting, "no one with whom I have talked in Berlin believes that Hitler is dead. They all think he 'got away'".[36]

The search for firm proof of death thus continued, although it was hampered by missing witnesses and mutual suspicion between the Allies. A memorandum produced by the Supreme Headquarters of the Allied Expeditionary Force at the end of July 1945 bemoaned the fact that "it is impossible to give any authoritative account of Hitler's last days since evidence is still accumulating. That which is already available is sometimes contradictory and incomplete and depends often on hearsay and conjecture. Much of the evidence, too, is in Russian hands".[37] The Americans, having captured Dr Blaschke themselves, proceeded to interrogate him about Hitler's dental history. Like Heusemann and Echtmann before him, Blaschke was able to recreate detailed descriptions and diagrams of the treatment he had performed on the Nazi leader—yet Allied investigators were hampered by the fact they had no post-mortem evidence to compare this to; Hitler's alleged jawbone and teeth were now archived in Moscow and the Soviets showed no signs of being willing to share this evidence.

To circumvent the lack of medical proof, the British and the Americans launched an extensive and time-consuming hunt for as many potential bunker eyewitnesses as possible. By the end of the process, Hugh Trevor-Roper was able to piece together accounts from secretaries Elsa Krüger and Traudl Junge who independently reported that Hitler had shot himself; Hitler Youth leader Artur Axmann who inspected the bodies and confirmed a bullet wound to Hitler's right temple; guard Erich Mansfeld who witnessed the removal of a body wrapped in a blanket; tailor Willi Otto Müller who saw five men carrying petrol on the evening of 30 April 1945; and the aforementioned Karnau who recognised the bodies as they were set on fire. The evidence, he noted,

> is not complete, but it is positive, circumstantial, consistent and independent…It is considered quite impossible that the versions of the various eyewitnesses can represent a concerted cover story; they were all too busy planning their own safety to have been able or disposed to learn an elaborate charade which they could still maintain after five months of isolation from other and under detailed and persistent cross-examination.[38]

Soviet investigators, meanwhile, spent the spring of 1946 re-visiting the purported scene of Hitler's death. Samples were taken from the bloodstained sofa in Hitler's living quarters while further examination of the bomb crater unearthed what was immediately considered to be the missing fragment of Hitler's skull, complete with apparent bullet hole. Once again, though, there was a refusal to make any definitive public statement on Hitler's death and, in the absence of any forensic proof of death, the Allies continued to be inundated with stories that Hitler and Eva Braun had escaped the bunker altogether. Letters were received from all over Germany, describing supposed sightings of the former leader, or promising to divulge important "facts" about his fate. Some accounts had them fleeing by plane to Denmark and thence to Argentina by submarine.[39] Others had them relocating to Munich, Hanover, or Hamburg, living under assumed names and the effects of plastic surgery. In September 1945, for example, the Hamburg story gained particular momentum through a series of sensational articles in the international media. Dr Karl Maron, Deputy Bürgermeister in East Berlin, inflamed matters by stating that he was "firmly convinced" that Hitler was still alive, and sea patrols began a search for the mahogany yacht believed to have conveyed the couple to safety. The British, who occupied this part of the country, were compelled to investigate these allegations, if only to be able to discredit them. A handwritten memo in the Foreign Office archives reveals the private sense that it was all "sheer poppycock". One commentator noted succinctly that the so-called "plastic operation" that had "changed Hitler's appearance" was probably carried out with a service revolver in the Führer bunker.[40]

The fact that such speculation existed owes much to the secrecy and contradictory messages disseminated by the Soviets about the forensic evidence in the summer of 1945. However, it can also be traced back to the sheer chaos in Germany during the final days of the Second World War. With the lines of communication broken, no clear political leadership, and the increasing threat posed by the advancing Red Army, everything had been in disarray, enabling rumours to spread like wildfire. Even Dönitz's official announcement of Hitler's death was experienced differently in different parts of the country. In the north, where Dönitz was trying to establish his provisional capital, the radio station had prefaced the broadcast with three warnings that "grave and important" news was about to be revealed, together with the playing of sombre music. It then held a three-minute silence in honour of the deceased. Consequently, the broadcast was rendered an event on North German radio. Listeners in the south, however, missed all of this. As the country teetered on the edge of collapse, many radio stations and other parts of the Nazi propaganda machinery had already fallen into Allied hands, reducing the Party's ability to disseminate a clear, uniform message. It was an hour and a half later that southern stations finally issued the news that Hitler was dead. Their audiences had not been prepared for this announcement as well as their northern counterparts; indeed, relatively light and cheerful music had been played up until midnight.[41] The timing of Hitler's death thus became fluid in the public imagination. The lack of a "proper" send-off on some radio stations may also have made it easier for people to doubt the accuracy of the reports.

What purpose did the survival stories serve, though? In part, documenting supposed sightings of Hitler may have simply been a form of attention-seeking, or even a deliberate attempt to stir up confusion between the Allies. It might also be argued that the rumour-mongers, having been denied any opportunity to mourn their leader, view his body, visit his final resting place or disseminate his image, were rebelling against the Allied "containment" of Hitler's death. Supposed sightings of Hitler and Braun enabled people to question the veracity of Allied pronouncements and imagine their own conclusion to the regime, regaining some element of control over the narrative. Alternatively, the very fact that people were volunteering "information" on Hitler's whereabouts to the authorities may be indicative of a desire to wreak revenge on the man held responsible for their current state of affairs, a hope that Hitler might yet be discovered and brought to justice for the damage he had inflicted upon the country. However, as Allied investigations focused on following potential leads to Hitler, rather than the characters of those making the sightings or spreading the rumours, we do not have the sufficient data to fully understand the motivations of these individuals.

That survival stories continue to emerge in the twenty-first century owes much to an enduring popular fascination with the Third Reich and the knowledge that other Nazis, such as Adolf Eichmann, did indeed manage to escape to far-flung locations after the war. More significant, though, is the fact that there remains some reasonable doubt about the thoroughness of the Soviet autopsies and the identification of the few body parts that have been retained since the exhumation of the Reich Chancellery gardens. In 2000, the skull fragment that had been retrieved in 1946 was "rediscovered" in the Russian archives and placed on public display in Moscow, generating a whole new wave of interest in the circumstances surrounding Hitler's death. In 2009, however, DNA analysis conducted by researchers at the University of Connecticut revealed that the fragment actually belonged to a woman under the age of 40, a result that immediately stirred up new conspiracy theories that rejected the narrative of Hitler's suicide in the bunker.[42]

The controversy surrounding the death of Adolf Hitler, then, shows no sign of abating. For the Allies operating immediately after the war, the aim was simple: find conclusive proof of the Nazi leader's death so that Nazism itself could be rendered truly dead. The western Allies, in particular, were all too aware that a lack of evidence could foster martyrdom myths, or fuel belief in Hitler's continued existence, thereby encouraging people to cling to the tenets of his ideology and fight on. A definitive end to the matter was considered not just desirable, but also achievable. An American cartoon published on 2 May 1945, the day after Dönitz's official announcement of the Führer's death, depicted a swastika draped body being removed from the ravaged Berlin landscape and asked whether this constituted "the end of the road".[43] Similar, if fleeting, optimism was expressed amid the initial confirmation that the charred remains discovered by the Soviets matched the available dental evidence for Hitler and, in 1956, there was renewed hope for closure when the district court in Berchtesgaden formally declared Hitler deceased and placed the death certificate on public display.[44] Hitler's "death" has thus occurred at multiple junctures. It is the failure, however, to unite legal, forensic and anecdotal proof of his demise that has enabled alternative versions of Hitler's fate to endure and keep him very much alive in the public imagination for all this time.

Notes

1. Thomas W. Laqueur, "The Deep Time of the Dead", *Social Research*, 78 (2011), p. 802.
2. For the most recent of these see Simon Dunstan and Gerrard Williams, *Grey Wolf: The Escape of Adolf Hitler* (New York, 2011); Paul Nelson dir., *Conspiracy* (2015); "Hola Hitler! Ex CIA Agent Claims Nazis Leader Faked His Death and Flew to Tenerife before Escaping to Argentina on a U-Boat", *Daily Mail*, 8 January 2016. The most recent claim—that Hitler fled to Tenerife—was printed across other British tabloids, including *The Sun*, *Daily Mirror* and *Daily Express*. The popular German press also has a tendency to relay such stories, but they usually stress their origins in the foreign media. See, for example, "Hitler Konnte Fliehen—Sollen FBI-Akten Beweisen", *Die Welt*, 7 October 2015; "Hat Hitler den Krieg Überlebt?", *Bild*, 7 October 2015.
3. Hugh Trevor-Roper, *The Last Days of Hitler* (London, 1947).
4. Lev Bezymenski, *The Death of Adolf Hitler: Unknown Documents from Soviet Archives* (London, 1968).
5. Ada Petrova and Peter Watson, *The Death of Hitler: The Final Words from Russia's Secret Archives* (London, 1995); Daniela Marchetti et al., "The Death of Adolf Hitler—Forensic Aspects", *Journal of Forensic Sciences*, 50 (2005), pp. 1147–1153. Most recently, scholarly attention has shifted onto the role of Allied intelligence gathering. See Sarah K. Douglas, "The Search for Hitler: Hugh Trevor-Roper, Humphrey Searle and the Last Days of Adolf Hitler", *Journal of Military History*, 78:1 (2014), pp. 159–210; Luke Daly-Groves, "The Death of Adolf Hitler: British Intelligence, Soviet Accusations and Rumours of Survival", Unpublished dissertation, University of Central Lancashire (2015).
6. The notion of 1945 as a "Zero Hour" or *Stunde Null* for Germany has enjoyed some currency over the years, giving the post-war German states a fresh foundation on which to construct their identities and distance themselves from the recent, Nazi past. See, for example, Konrad Jarausch, "1945 and the Continuities of German History: Reflections on Memory, Historiography, and Politics". In Geoffrey J. Giles ed., *Stunde Null: The End and the Beginning Fifty Years Ago* (Washington, D.C., 1997), pp. 9–24.
7. There is a growing literature on Nazi death cults. See, for example, Jay Baird, *To Die for Germany: Heroes in the Nazi Pantheon* (Bloomington, Indiana, 1990); Sabine Behrenbeck, *Der Kult um die Toten Helden: Nationalsozialistische Mythen, Riten und Symbole 1923 bis 1945* (Cologne, 2011); Jesús Casquete, "Martyr Construction and the Politics of Death in National Socialism", *Totalitarian Movements and Political Religions*, 10:3, (2009), pp. 265–283; Peter Lambert, "Heroisation

and Demonisation in the Third Reich: The Consensus-Building Value of a Nazi Pantheon of Heroes", *Totalitarian Movements and Political Religions*, 8:3–4 (2007), pp. 523–546; Daniel Siemens, *The Making of a Nazi Hero: The Murder and Myth of Horst Wessel* (London, 2013).
8. "Zum 9 November 1942: Gedenktag für die Gefallenen der Bewegung", *Die Neue Gemeinschaft*, 8 September 1942. Author's emphasis.
9. Cited in Jeremy Noakes ed., *Nazism: A Documentary Reader, 1933–1945. Vol. 4: The German Home Front in World War II* (Exeter, 1998), pp. 624–626.
10. The term "social death" is frequently invoked within Holocaust studies to describe how Nazi policies of discrimination and segregation steadily rendered the German public indifferent to the treatment of the Jews; ostracism from mainstream society during the 1930s was the first step towards physical extermination during the Second World War. It is the contention of this chapter, though, that the term can offer valuable insights into the fate of Nazi perpetrators, a group hitherto overlooked within studies of post-war remembrance culture.
11. John Borneman ed., *Death of the Father: An Anthropology of the End in Political Authority* (New York, 2004), p. 2.
12. "Doenitz Announces Hitler's Death", 1 May 1945. Jewish Virtual Library. Accessed at: https://www.jewishvirtuallibrary.org/jsource/Holocaust/hitlerdeath.html. This statement was reproduced in the international media. See, for example, "Donitz [sic] as Head of State", *The Times*, 2 May 1945.
13. "Doenitz Announces Hitler's Death".
14. "Abschied von Hitler", *Hamburger Zeitung*, 2 May 1945.
15. FO371/46748: Review of the Foreign Press, The National Archives, Kew (hereafter TNA). Author's emphasis.
16. Witness testimonies from bunker staff and several leading Nazis agree that Hitler had expressed a desire to commit suicide from 22 April 1945. See WO208/3781, TNA.
17. FO371/46748: Roberts to Foreign Office, 2 May 1945, TNA.
18. FO371/46748: Harrison memorandum, 2 May 1945, TNA.
19. Meeting between Heinrich Himmler and Count Folke Bernadotte reported in FO371/46748: Washington to AMSSO, 2 May 1945, TNA.
20. FO371/46748: Harrison memorandum, 2 May 1945, TNA.
21. FO371/46748: Yarrow telegram, 4 May 1945, TNA. Eisenhower's statement was reprinted in the press. See, for example, "Hitler Met No Hero's Death, States 'Ike'; Was Dying of Brain Illness over Week Ago", *Ottawa Citizen*, 2 May 1945; "Eisenhower's Exposure of New 'Fuehrer'", *Daily Telegraph*, 3 May 1945.
22. By mid-May 1945, Himmler's own fate had become the subject of some discussion. Responding to a question in the House of Commons,

Winston Churchill declared, "I expect he will turn up somewhere in this world or the next and will be dealt with by the appropriate local authorities. The latter would be more convenient to His Majesty's Government". See FO371/46748: PMQs, TNA.
23. FO371/46748: AFHQ to German Department, 7 June 1945, TNA.
24. "Hitler's Doctor Denies Führer was Ill", *Daily Telegraph*, 7 May 1945.
25. FO371/46748: PMQ by Major Anstruther-Gray, 15 May 1945, TNA.
26. Cited in V. Vinogradov et al., *Hitler's Death: Russia's Last Great Secret from the Files of the KGB* (London, 2005), p. 262.
27. FO371/46748: Roberts to Foreign Office, 6 May 1945, TNA.
28. The Soviet autopsy report was first published in 1968 in Bezymenski, *The Death of Adolf Hitler*, pp. 44–51. Excerpts were later reproduced in Marchetti et al., "The Death of Adolf Hitler", pp. 1147–1148.
29. Marchetti et al., "The Death of Adolf Hitler", p. 1148.
30. Cited in Vinogradov et al., *Hitler's Death*, p. 108.
31. See, for example, "Reported Finding of Hitler's Body", *The Times*, 7 June 1945.
32. "Hitler's Last Hours", *The Times*, 21 June 1945.
33. Cited in Petrova and Watson, *The Death of Hitler*, p. 44.
34. "Body Russians Found Was Not Hitler's", *Daily Telegraph*, 5 July 1945. See also "The Body Outside Hitler's Shelter Was Not His", *News Chronicle*, 5 July 1945.
35. "Jawbones identified as Hitler's", *The Times*, 9 July 1945.
36. "Hitler Still Alive Says Moscow", *Daily Herald*, 5 July 1945.
37. FO371/46749: Memorandum by the Supreme Allied Expeditionary Force, 30 July 1945, TNA.
38. WO208/3781: "The Death of Hitler" (undated), TNA.
39. The Denmark story gained particular traction towards the end of 1947 when Peter Baumgart, a former Luftwaffe pilot, was named as the person who had piloted the Hitlers to Denmark via Magdeburg.
40. FO371/46749: Annotation on telegram from Roberts to Foreign Office, 12 September 1945, TNA.
41. FO371/46748: Review of the Foreign Press, 22 May 1945, TNA.
42. See, for example, "Tests on Skull Fragment Cast Doubt on Adolf Hitler Suicide Story", *The Guardian*, 27 September 2009. The results of the DNA analysis were also featured in *Mystery Quest: Hitler's Escape* (History Channel), broadcast date 16 September 2009.
43. "The End of the Road?", *Providence Journal*, 2 May 1945.
44. "The Death Certificate of Adolf Hitler", 25 October 1956, Associated Press Archive. Accessed at, http://www.aparchive.com/metadata/youtube/1bc3d33e28bb42fe8ad521c890c644d3.

Author Biography

Caroline Sharples is Senior Lecturer in Modern European History at the University of Roehampton. Her research interests encompass postwar German commemorative cultures, war crimes trials and representations of the Holocaust. She is the author of *West Germans and the Nazi Legacy* (Routledge, 2012), *Postwar Germany and the Holocaust* (Bloomsbury, 2015) and co-editor, with Olaf Jensen, of *Britain and the Holocaust: Remembering and Representing War and Genocide* (Palgrave Macmillan, 2013).

Open Access This chapter is licensed under the terms of the Creative Commons Attribution 4.0 International License (http://creativecommons.org/licenses/by/4.0/), which permits use, sharing, adaptation, distribution and reproduction in any medium or format, as long as you give appropriate credit to the original author(s) and the source, provide a link to the Creative Commons license and indicate if changes were made.

The images or other third party material in this chapter are included in the chapter's Creative Commons license, unless indicated otherwise in a credit line to the material. If material is not included in the chapter's Creative Commons license and your intended use is not permitted by statutory regulation or exceeds the permitted use, you will need to obtain permission directly from the copyright holder.

CHAPTER 7

Death's Impossible Date

Douglas J. Davies

INTRODUCTION

"When is death?" is an apparently simple question; this chapter argues that death has an impossible date. This rather enigmatic response is teased out in three sections, each of which briefly surveys different ways that the question can be tackled: Chronological Precision, Life-course Narratives, and Existential Anticipation. The first presents some cultural measures on the timing of death, including a Mormon case study; the second surveys perspectives from anthropology and bereavement studies; the third offers further anthropological perspectives taken in a more existential direction.

CHRONOLOGICAL PRECISION: CULTURAL MEASURES OF DEATH

Death Certification

In British society, the question "When is death?" is primarily answered on a medical death certificate where we find a date and place of death, but not the time of death as such. A medical doctor marks this event

---

D.J. Davies (✉)
Durham University, Durham, UK
e-mail: douglas.davies@durham.ac.uk

and becomes the agent of society in accounting for death. The certificate then becomes a valuable document in how people manage the numerous legalities involving the dead person's estate. Assessing death after the event is, of course, a more complicated medical task and may, if circumstances demand it, require forensic pathology, or even a police investigation of the circumstances surrounding death. In some cases, the dating of a death is an important factor in deciding whether a standard death certificate is used (as for most people), or whether a Certificate of Stillbirth is to be used for those who have passed 24 weeks of gestation, but did "not breathe or show any other signs of life after being completely expelled from its mother". If a child was born alive and lived for at least 28 days before dying, then a Neonatal Death Certificate is required. While such timings of the death of a foetus or live-born child are important for legal and medical purposes, they can carry a different significance for parents who may well have experienced an increasing sense of having been "parents" or "parents-elect" during the ongoing nine months of pregnancy—a status enhanced by scans and photographs of their child in utero.

In British society today, people experiencing a stillbirth, or the death of a baby within that four-week period of birth, encounter a very different parental life-course narrative than the narratives which predominated in the past. Notably, in the case of stillbirths, the parents may well wish to claim their right, as it were, to parenthood. In contrast to the days when a neonatal death might be treated as a "medical waste" issue, with the mother having no contact with the "child", there are now bereavement support networks available and some parents choose to have photographs taken with the "child". The emergence of an appropriate funeral for such births has become one social marker of "lives" whose only biological forms are intra-uterine, but which are accorded social lives through commemoration. Hospital chaplains also sometimes baptise such births, even though in the theological terms of mainstream Christianity, one can only baptise biologically living people (though, as we will see, there is an exception to this in Mormonism). These neonatal death contexts make the question "when is death?" very problematic.

Medically Sustained Life

In our modern, medically-advanced society, prematurely born and terminally ill babies can be nurtured through life-support systems. This also applies to severely ill adults and to aged individuals in ways that raise complex medical, ethical, religious, and philosophical issues over the question "when is death?". There are many people who would have died "naturally" if they had not been treated "culturally" through medical intervention. The distinctive concepts and processes of, for example, "brain-death" and organ donating, clearly raise question marks over whether a "person" is "alive" or dead, or can be given a new lease (and status) of life through replacement organs. Medical practitioners frequently have to make difficult decisions about keeping people alive. For instance, should the victim of a serious accident be kept ventilated and on life-support, even if the victim's brain injuries make any realistic hope of recovery an unlikely prospect? In this scenario, the victims' organs could play a key role in the recovery of otherwise terminally ill people. This, in itself, raises the question: when is death for distinctive body parts, given that hearts may survive for four hours or so and kidneys might be sustained for 36 hours?[1] What forms and timings does life take when it is sustained by the lively organs from a recently "dead" person?

The Departure of the Soul

The theme of animation must also be considered because of the widespread popular idea of death as the absence of the soul, spirit, or life force from the body. As a near-universal perspective on life, human beings interpret death as the removal or loss of vitality, whether in natural philosophy, natural medicine, or natural religion. Given the power of what cognitive anthropologists and others call the "animacy principal"— our hardwired capacity to sense agency or "life" in things—people have tended to assume that death occurs when the body no longer breathes, when the breath of life departs. Breath has, for millennia and in many parts of the world, symbolised life, while its absence marked a person's death. The reification of bodily life in breath, and then in soul, has meant people have thought of the soul as having an existence all its own, not least outside, or beyond the body.

Yet, life's departure is seldom seen as instantaneous, in that in life a "soul" may linger around the body; be offended if relatives mourn too much or do not mourn enough; and move on to some new identity in another domain. Today, in Britain, for example, Muslims are buried in such a way that they can sit up shortly after burial to answer key questions put to them by the post-mortem visiting angels. Or again, traditional Rabbinic Judaism echoes the idea of a lingering spirit, alert to the behaviour of surviving family members. This should not be provoked by bodily activities, including sexual intercourse, in which that spirit can no longer participate.[2] These and many other cases suggest that a body/soul distinction is a means by which death can be thought of as impossible, despite the presence of a corpse.

This impossibility comes from the idea that there is no mortality as far as vitality-force is concerned; "persons" do not die, they change. Belief in the existence of an after-life can be found in countless religious and cosmological traditions, such as the karma-related transmigrations in Indian-derived worldviews and the journey through judgement to paradise in middle-eastern traditions. Even in secular societies, where many people think of death as the complete end of a person's vitality-existence, memorialising behaviours mean that the memory of the dead lives on and influences the behaviour of the living. The "when" of death, in other words, is hard to calculate when the presence of a person is obviously not erased when he/she dies.

Mormon Death

My previous work on Mormonism suggests that timing death is something that involves the manifestation of "life" in a series of its modes.[3] In Mormonism, each person is thought of as pre-existing as an entity known as "intelligence". That intelligence then comes under the influence of a more advanced intelligence known as Heavenly Father who transformed or engendered "intelligence" into a spirit-child. This spirit then comes into existence in a kind of pre-mortal heavenly domain where it joins with an earthly body to create a "soul". This Mormon terminology is often seen as counter-intuitive by other Christians because it speaks definitively of spirit plus body producing a "soul", rather than speaking loosely of a soul and body as constituting a human being. For Latter-day Saints, however, death is thought of as the spirit leaving the body (meaning the "soul" no longer exists). The soul then passes to the spirit world and the body to the grave until the day of resurrection. At this time, the spirit will engage with a resurrected body in a transformed unity that will now be judged and move into one of a whole series of post-mortal domains, known as kingdoms or degrees of glory.

So, for Mormons, the question "when is death?" is answered on the one hand by the separation of spirit from body when the "dead" body is washed and dressed in sacred clothing before being buried (burial being more usual than cremation in Mormon culture). On the other hand, the spirit now exists elsewhere and will, at a future date, be reunited with its transformed body and move onwards into cosmic glories. However, even this is but a partial picture of life and death, for the devoted Mormon will have spent a significant amount of earthly time performing rituals within the distinctive sacred space of the temple. These are not the ordinary churches found in most towns, but the one or two temples present in most nations into which only accredited and approved church members can gain access. In the intense ritual activity of these Mormon Temples, the living are baptised on behalf of the dead, for whom they have collected family history. As spirits, the dead await the living and avail themselves of the dynamic opportunities available to them once vicarious baptism and other key rites, such as ordination and marriage, are conducted on their behalf. Temple and genealogical work is frequently said to bring the living and those in the spirit world close together, signified by the saying that in the temple the "veil is very thin" between this world and the next (a metaphor materialised by the literal veils which separate different qualities of existence in the temple).

The Mortality Paradox

The Mormon case illustrates the fact that "the dead" are, in a sense, alive for them. In the after-life they wait for the living to engage in the ritual activity that offers an enhanced form of eternal-cosmic life after the God-given resurrection of all. The extensive genealogical and ritual work of Mormons on behalf of the dead keep the living in mind, not in the simple memorialist sense of a family tree, but in a pro-active sense of creative endeavour on behalf of forebears. The Mormon case also pinpoints what we might call the mortality paradox. This feature of many religious traditions sets an emotional awareness of loss against a belief in the continuing existence of the dead, albeit in a changed state, "place", or condition. Grief, the result of this dissonance, implies that there is no precise timing to death, only a set of timings to different states of being.

In other Christian traditions, the life of the "dead" person has been variously described in terms of being asleep, or in some post-mortal intermediate state prior to its final destiny with God. Inspired by the early work of Sir James Frazer, anthropological accounts reveal that the mortality paradox is also widespread in non-western societies.[4] For instance, traditional Indian-originating views on birth, dying, death, and afterlife rites speak about the animating force that comes to the foetus in utero via its cranial sutures, and which departs when the skull is cracked on the funeral pyre. The very notion of transmigration of the life-force under the dynamic moral schemes of *karma* attests to its non-death and samsaric processing from agent to agent over expanses of "time".[5]

In many contemporary cultural domains, traditional worldviews have given way to a secular ideology where the mortality paradox takes quite a different form. Here, the "when is death?" question becomes subject to medical judgement and to ethical issues of identity and the dignity of the person's body. In the context of euthanasia, organ donation, and terminal illnesses, people on ethics committees now play roles in the timing of death and declarations of "social death".

Life-Course Narratives

A high proportion of our lives are taken up by talking about the lives of others: from family and friends to strangers and celebrities, humans generate social narratives by speaking, gossiping, writing, consuming media, praying, and so on. One way that people engage with others who are deceased is through the notion of a "continuous present". In his famed essay on funerary rites, Robert Hertz argued that "[s]ociety imparts its own character of permanence to the individuals who compose it: because it feels itself immortal and wants to be so, it cannot normally believe that its members ... in whom it incarnates itself should die".[6]

Despite all the criticism that can be laid against Hertz's Durkheimian reification of "Society", he offers a powerful image of how individuals are entangled with society; this idea undergirds practically all social theory concerning identity but also, by extension, theories of grief. Immortality, according to Hertz, is not an absence of death: rather death is timed according to a sense of value derived from the experience of relationships, not in terms of chronology as such. When, for example, Scots toast "the immortal memory" of Robert Burns at innumerable Burns Night Suppers across the world, they are referring to his cultural value and its identity-generating capacity, rather than his timelessness. At the family level, too, this happens when many ordinary families possess a sense of three or four generations from a set of ancestors, but can still speak of their dead as inhabiting some "timeless" realm. Hertz, then, brings our analysis of death and time into the world of social relationships, identity, and embodiment. What he says of death raises similar questions about life, for societies have differed over the age at which they accord some firm identity to "first-life" (i.e., to a child), especially in cultures where infant mortality was common and where, we might say, no cultural energy would be invested in an infant until it looked as though it would become a valuable social commodity. This is not to say that mothers, or others, might not grieve over an infant's death, but it is to note the attitudes of social networks into which infants are only more clearly drawn over time. We have already indicated the significance of medical technology in contemporary Britain, enabling pre-birth images to ascribe the foetus its own kind of social personhood. In other words, the question "when is death?" is now haunted by its double: "when is life?".

"Grief Mean Time"

What, then, of bereavement in life-narratives? One way to think about the timing of death is through the notion of a "Grief Mean Time" (GMT), a standard orientation point of loss in time that is nonetheless experienced by people in different ways. Here care is needed lest the mind fly too rapidly to Elisabeth Kűbler-Ross, whose stage theory of bereavement was adopted as a chart-index with its "denial, anger, bargaining, depression, and acceptance" phases.[7] While extensive criticism of the schematized version of this perspective reveals an absence of a solid empirical base, its popularity reflects how important a narrative journey is for the bereaved.[8] Another 1960s volume, Geoffrey Gorer's *Death, Grief, and Mourning* (1965), described grief as a "long-lasting psychological process with physiological overtones and symptoms", especially "disturbances of sleep and weight loss". For Gorer, this GMT begins before death, as in cases of incurable illness when "a great deal of mourning may take place during this period so that the eventual death is felt emotionally, as well as intellectually, to be a release".[9] The question "when is death?" thus becomes "when is grief?"—a question that pivots on shared understandings of the staged timings of bereavement. Grief can commence when a person dies socially, long before biological death occurs. For instance, the issues of identity-loss and recognition in Alzheimer's disease highlight the bio-cultural nature of death, suggesting an increasing arc that may plateau out in the social path to death, but also falls as biological life is maintained whilst a person's social significance declines.

In terms of theories of grief, Gorer's encompasses both the attachment-loss theory of grief and of the "continuing bonds" tradition. The qualitative nature of time in emotions is also evident in Peter Marris's

1956 study of 72 working-class widows in the East End of London, *Widows and Their Families*. This offered a "three stages of loss of contact with reality" that moved from an "initial period of shock", to "violent grief and disorganization" (of about 6–12 weeks in Britain), and a final "longer period of reorganization".[10] These stage-theories of grief suggest that death's "mean time" is experienced as a series of emotional patterns that may come and go or re-pattern themselves in the time before and after the biological death of a loved one.

Experiencing the Dead

It is important to recall that we experience the presence of the dead every day. As phenomenologists frequently argue, we live in a world of multiple realities constituted by our interactions with places, minds, and bodies. The dead feature in these interactions in popular concepts like nostalgia, homecoming, and ghost-seeing. Indeed, in a major research project published in 1995, 1603 people were interviewed in their homes and asked if they had experienced a sense of the presence of a dead person after they had died.[11] A sizable minority of people—roughly 35%—said that they did have such an experience, as Table 7.1 indicates.

The relationship to the deceased person is given in Table 7.2 (the reason why the total percentage comes to 40% and not 35% as above reflects the fact that a few experienced the presence of more than one relation). The experiences were said to have taken place mostly at home or at a relative's home (approx. 54%). Other contexts for the experience included, in a dream (4%), when ill or in hospital (3%), in association with a pen (3%), at a Spiritualist meeting (2%), at a graveside (1.6%), or when driving a car (1.4%).

While these analytical categories are far from perfect, they indicate an aspect of life that is easily ignored by outsiders, but can be profound for the bereaved individual. The fact that 23 individuals said they were not sure if they had had such an experience suggests an inability among some people to establish hard lines between memories and emotional awareness. Indeed, this is a hard thing to do because, apart from palpably

Table 7.1 Experienced presence of the dead	*Often*	*Occasionally*	*Just once*	*Rarely*
	8.5	13.6	7.3	5.9

Table 7.2 Relation—experience

Relation	Incidence
Parent	15.4
Grandparent	10.3
Spouse	5.0
Sibling	2.2
Child	1.1
Other kin	3.6
Friend	1.7
Non-kin	0.7

sensing presences in a supernatural or unexplained way, every day millions of Britons look at photographs of the dead or think about people they have lost (see Table 7.2).

Liturgical Time and Death: Triadic Moments

In a separate study focused on Anglican churchgoers, it was found that approximately 36% of people who attend the Holy Communion Service said this helped give them a sense of presence to their dead.[12] That, we might, suggest is not an unexpected finding for a formal ritual that normatively names the dead, prays for the departed, and speaks of a united community on earth and heaven. The liturgy of the Eucharist—one of the most long-lived and cross-culturally widespread of all human ritual behaviour—is especially important for the "when is death?" question since its very nature is embedded in the notion of the historical death of Christ, coupled with belief in his resurrection from the dead. Those dynamics then frame the life, death, and promised eternal life of believers participating in the rite of Holy Communion. Communicants participate in the complex ritual symbolism of the Eucharist every time they eat and drink the body and blood of Christ. These are elements that stand, at one and the same time, for the death of the Saviour whose "living presence" frames the devotional piety of Christians. Given that the saintly dead, as well as those who might have died recently, are often named during the Eucharist, and given that churches often also conduct the funerals of the dead at the same time as the Eucharist, it is not surprising that Christian churches foster the mortality paradox.

In her research on bereavement, Christine Valentine recounts interview situations in which, as one person talks about the dead, a period would arise in which it seemed as though the deceased person was actively present to interviewee and interviewer.[13] This kind of third-party "presence" is, I suspect, likely to be familiar to many who engage in pastoral work with the bereaved. Here, narrative comes into its own as it generates a form of momentary transcendence. The story of a dead person told to others, in other words, creates a presence from the absence. Talking to the bereaved reveals a triadic relationship: an interviewer becomes aware of the intensity of your relationship with someone who lives in your life-narrative in such a way that the third-party assumes a kind of invisible social presence. As for narrative accounts of the dead, we have already mentioned that the time of death may be largely ignored in medical-legal certification, but such timing is often quite different in life-course narrative where relatives detail the story of a death, of their own presence or absence, and of the time of day. When and where their relative died is significant: perhaps in the early hours of the morning when they were at the bedside, perhaps the fact that they managed to get there in time, or arrived too late. In such contexts, time matters more than dates, and in this death-bed sense, death is usually marked with specificity. Moreover, the mortality paradox seldom seems relevant, or emerges in a distinctive way, as relatives comment on the fact that the body that now lies there is no longer the "person" they knew and loved. In the moment, death's all too evident date also marks life's departure, and it is often through such moments that popular beliefs about the mobility of the soul seem self-evident.

Body-Recall

A related form of sensed presence comes in what might be described as "body-recall". By this, I refer to the experience of seeing in one's physical form something one recalls from their deceased parents' bodies. This could be an ageing face, a mode of walking, the sound of a cough, or indeed any number of things that remind us of our dead parent or blood relation (I stress such consanguineal kin precisely because of genes and body shape). At this moment of body-recall, the memory of the dead is activated in my living body and may, or may not, be a reminder that I am aging and will die. Such experiences are linked to another distinctive body technique, where the recall of another person's bodily deportment is evoked by my own body. In one striking case, a retired Anglican bishop who was interviewed for a study, indicated a pen that happened to be at hand and said that whenever he wrote with it he recalled his father in the act of writing.[14] Perhaps the most interesting factor of such "memories of the flesh" are that they require both the death of the parent, and the aging of the child to be close to the adult age of the dead parent.

Old Graves

In the final survey of this section, I return to the Report on Popular British Attitudes, mentioned above. In this study, 1603 individuals were specifically asked: "What do you think would be a respectable time lapse before an old grave might be used for new burials by a different family?" Respondents were free to give any number of years they wished, and of the total sample, only a slight majority 55% (or 875 individuals) did feel able to give an answer. Among these, a 50-year period attracted approximately 20% of support, and a 100-year period 39%.[15] Another question asked for reasons why there should be a time lapse before a grave is reused. The responses included: that there would be no one left to tend a grave (44%); that the dead should rest in peace and not be disturbed (24%); that the body would be decomposed (9%); and that time is needed to grieve (7%). If not providing an absolute answer to the "when is death?" question, these responses do provide a sense of how long it might take for people to redefine their relationships with bodies in graves. Here, images of continuing bonds combine with processes of separating from the dead. These responses show that "when?" is marked not by a specific date, but by change over time and, in that sense, we are reminded again of the narrative nature of death within its cultural frame.

EXISTENTIAL ANTICIPATION

In this section, I survey the theoretical ideas of dual sovereignty and paradigmatic scenes.[16] The first idea, dual sovereignty, concerns forms of authority in human life that are balanced between jural (legal) authority and mystical authority. In terms of death, mystical authority ranges from the ancestral capacity to bless or curse descendants to modern society's concern with ideas of respect, dignity, and a "good send off" for the dead. Its complement, jural authority, also covers a spectrum of existence including what ecclesial or civic authorities allow to be inscribed on gravestones; legal decisions covering the duty of care; and issues of harm, murder, and suicide. In terms of the "when is death?" question, these dual forms of authority are non-controversial together in situations where, for instance, medical certification of death complements a family's desired funerary provision. Problems emerge if and when, for example, a religious group might want a rapid burial of a dead body (meeting its mystical authority), while the state requires much more time for full authorisation. Another flashpoint concerns assisted dying. If I think I have the right to choose assisted dying, but the law in Britain prevents it, the dual sovereignty balance is upset and a certain social disease ensues. This problematic ethical, medical, and religious topic leads immediately to the complementary theoretical issue of the paradigmatic scene. In a media-flooded world where photographic images carry powerful significance, paradigmatic scenes (e.g., icons, art, and statuary) sustain the core messages of religious traditions, not least in terms of death transcendence. In terms of the "when is death?" question, people imagine an apartment in Switzerland where a person goes to die, or the image of an old-age home, a television and a circle of arm-chaired and relatively inattentive viewers. Each is a paradigmatic scene capturing core values and reflecting the demise of vitality; in the one, a life is intentionally ended because it is felt no longer to be a flourishing; in the other, life seems interminably protracted and lacking in vitality.

By contrast, another paradigmatic scene, one framed by a balanced dynamic of dual sovereignty, is that of the "woodland", "natural", "green", or "ecological" burial. Emerging in the UK in the mid 1990s, these kinds of burials now occur in about as many sites as there are crematoria. Here, people generate a paradigmatic scene in a kind of hospitable garden-centre-like locale where the body is thought of as passing into the natural environment.[17] Such an anticipated context resets the "when" of "when is death?" as a "where" of an anticipated merger of self and world. This kind of shift in discourse is not unique in relation to death in Britain: it occurred both in the late nineteenth century with the innovation of modern cremation and in the 1970s in terms of how cremated remains were dealt with.

CONCLUSION

"When is death?" strikes me as a question that has something of the character of a *koan* about it. The *koan* is a Zen Buddhist presentation of a problem "insoluble by, and nonsensical to, the intellect". It is aimed at "breaking through intellectual limitations" to produce "a flash of insight".[18] For example, the popularised *koan* "What is the sound of one hand clapping?" temptingly provokes a logically consistent "answer", rather than being taken as a verbal form aimed at catalyzing a different form of understanding.[19] One such shift in understanding might involve the difference between the dual classification of things (as in "two hands"), and the singularity of things ("one hand"), all as part of a reflective-meditative tradition of self-understanding. Committing to such word-play practice can, experientially, shake the easy confidence that the everyday use of language confers. When reading the thoughts of philosophers on death, I am often possessed of an eager anticipation that soon passes into disappointment. It is as though I am sure that each sentence will lead to another in a logical flow that will end in satisfaction: the conclusive key will open the safe to reveal the desired sight. However, while sentences help set the scene, and offer some glimpsed novel vista, the horizon remains shrouded in mist.

If "when is death?" stands as a question grounded in the ideas of both mortality and time, then my title, "death's impossible date" stands as something of a different logical type. This is why no answer can be given to the question, and why the *koan* motif provides both a constraint on answering and a freedom not to answer. It provokes a shift in understanding. So, I conclude with a formulation rather than an answer: death attracts some emotional affect to render it as a value; for some, this value enters into a person's sense of identity and thus becomes a belief; and for some others this belief constitutes a sense of identity and becomes a religious belief. Death has an impossible date because the "when" of death is not coeval with "the time of not being". Death can mean the beginning of a sensed presence of the dead, or of an eternal God with whom a sensed affinity seems to guarantee one's own immortality. Experience counts, and behavioural acts frequently foster experience. This is where philosophy is at a disadvantage since its westernised manifestation knows no ritual but the lecture and seminar. Theology, meanwhile, possesses the advantage of being able to ritualise its utterances in liturgy or in private prayers. These may prompt an insightful awareness of mortality and vitality in "death's impossible date".

Notes

1. Nora Mochado, "Organ and Tissue Donation and Transplantation". In Clifton D. Bryant and Dennis Peck eds., *Encyclopedia of Death and the Human Experience* (Los Angeles and London, 2009), p. 789.

2. See David Kraemer, *The Meanings of Death in Rabbinic Judaism* (London, 2000).
3. See Douglas J. Davies, *Introduction to Mormonism* (Cambridge, 2003).
4. See James George Frazer, *The Belief in Immortality and the Worship of the Dead, Vol. I. The Belief among the Aborigines of Australia, The Torres Straight Islands, New Guinea and Melanesia* (London, 1913); James George Frazer, *The Belief in Immortality and the Worship of the Dead, Vol. II. The Belief among the Polynesians* (London, 1922); James George Frazer, *The Belief in Immortality and the Worship of the Dead, Vol. III. The Belief among the Micronesians* (London, 1924); Douglas J. Davies, "Death, Immortality, and Sir James Frazer", *Mortality*, 13 (2008), pp. 287–296.
5. See Jonathan Parry, *Death in Banaras* (Cambridge, 1994).
6. Robert Hertz, "A Contribution to the Study of the Collective Representation of Death". In R. Needham and C. Needham eds., *Death and the Right Hand* (New York, 1960), p. 77.
7. See Elisabeth Kübler-Ross, *On Death and Dying* (New York, 1989).
8. See Marc P.H.D. Cleiren, *Adaptation after Bereavement: A Comparative Study of the Aftermath of Death from Suicide, Traffic Accident, and Illness for Next of Kin* (Leiden, 1991); Charles A. Corr, "Coping with Dying: Lessons That We Should and Should Not Learn from the Work of Elisabeth Kübler-Ross", *Death Studies*, 17:1 (1993), pp. 69–83.
9. Geoffrey Gorer, *Death, Grief and Mourning* (London, 1965), pp. 53, 68.
10. Cited in Douglas J. Davies, *Mors Britannica: Lifestyle and Deathstyle in Britain Today* (Oxford, 2015), p. 216.
11. See Douglas J. Davies and Alastair Shaw, *Reusing Old Graves: A Report on Popular British Attitudes* (Crayford, Kent, 1995).
12. See Douglas J. Davies, Charles Watkins, and Michael Winter, *Church and Religion in Rural England* (Edinburgh, 1991).
13. See Christine Valentine, *Bereavement Narratives: Continuing Bonds in the Twentieth Century* (London and New York, 2008).
14. See Douglas J. Davies and Mathew Guest, *Bishops, Wives and Children: Spiritual Capital Across the Generations* (Aldershot, 2007).

15. Davies and Shaw, *Reusing Old Graves*, pp. 42–43.
16. See Davies, *Mors Britannica*; Rodney Needham, *Reconnaissances* (Toronto, 1980); Rodney Needham, *Circumstantial Deliveries* (Berkeley and Los Angeles, 1981).
17. See Douglas J. Davies and Hannah Rumble, *Natural Burial* (London, 2012).
18. D. H. Smith, "Koan". In S. G. F. Brandon ed., *Dictionary of Comparative Religion* (London, 1970), p. 398.
19. John Bowker, "Koan". In John Bowker ed., *The Oxford Dictionary of World Religions* (Oxford, 1997), p. 552.

Author Biography

Douglas J. Davies trained in both anthropology and theology. He is Professor in the Study of Religion and Director of the Centre for Death and Life Studies at Durham University, a Fellow of the Academy of Social Sciences, and also a Fellow of the British Academy. His most recent monograph is *Mors Britannica: Lifestyle and Death-style in Britain Today* (Oxford University Press, 2015).

Open Access This chapter is licensed under the terms of the Creative Commons Attribution 4.0 International License (http://creativecommons.org/licenses/by/4.0/), which permits use, sharing, adaptation, distribution and reproduction in any medium or format, as long as you give appropriate credit to the original author(s) and the source, provide a link to the Creative Commons license and indicate if changes were made.

The images or other third party material in this chapter are included in the chapter's Creative Commons license, unless indicated otherwise in a credit line to the material. If material is not included in the chapter's Creative Commons license and your intended use is not permitted by statutory regulation or exceeds the permitted use, you will need to obtain permission directly from the copyright holder.

CHAPTER 8

The Legal Definition of Death and the Right to Life

Elizabeth Wicks

INTRODUCTION

The law engages with the issue of death in various manners and contexts. For example, criminal law prohibits killing; inheritance law regulates the redistribution of property after death; and medical law determines when a patient should receive life-sustaining treatment, as well as when a body's organs become available for transplantation purposes. The law provides regulation and clarity to the life-death boundary. It is greatly influenced, however, by clinical, social, and moral conceptions of death and dying. Indeed, the legal definition of death in the UK is merely the judicial application of the current medical definition of death. In this chapter, the relationship between the legal definition of death and the legal protection of a right to life in human rights law will be considered in order to provide some legal perspectives on the eternally challenging question of "when is death?"

E. Wicks (✉)
University of Leicester, University Rd, Leicester LE1 7RH, UK
e-mail: eaw19@leicester.ac.uk

The Legal Definition of Death

There is often perceived to be a clear distinction between life and death. Indeed, it is the most fundamental distinction in the experience of humanity. In reality, however, the line is blurred. In traditional biological understandings, death occurs when an individual ceases breathing or when his or her heart stops beating. Such cardio-respiratory failure is no longer an adequate conception of death, however, because advances in medical technology have enabled the restarting of a heart that has stopped beating, as well as artificial respiration to counter a cessation in breathing independently. The consequence of this is that a person who would once have been regarded as dead—one who is not breathing and/or whose heart is not beating—can now be revived. Death, in the sense of cardio-respiratory failure, has been conquered. And yet, death remains.

From a legal perspective, such ambiguity is unsatisfactory. The line between life and death must be differentiated for a variety of social and legal reasons. The availability of organs for transplant, rules of inheritance, criminal liability for causing death, and the need for disposal of the body are all issues necessitating a clear line between life and death. There needs to be a clear-cut, and unambiguous, definition of death within the law because legally we treat a dead body very differently from the way we treat a living person.

The law's response to the indeterminacy of cardio-respiratory failure has been a focus upon the death of the brain. A committee of the Harvard Medical School in 1968 first offered a set of criteria by which doctors could establish that a patient had suffered permanent loss of all brain functions. Significantly, the Committee also proposed those criteria as a diagnosis of death.[1] Subsequently the United States adopted a uniform model death law. The Uniform Determination of Death Act is now law in 36 US states and contains a split legal standard: both cardio-respiratory failure and whole-brain death are regarded as legal death in these jurisdictions. This does not, of course, mean that there are two different ways to die in the US today. Rather, it means that there are two different ways for doctors to determine that somebody has died. The brain death variation avoids any possibility of subsequent medical intervention; once a brain has died, there is no known treatment for revival.

The whole-brain death criteria has been somewhat modified within the UK where both the clinical and legal emphasis is on brain stem death. The brain stem's functions include responsibility for generating the capacity for consciousness and the respiratory centre. Significantly, it is also the part of the brain least effected by a lack of oxygen and; therefore, it can be assumed that if the brain stem is irreversibly destroyed by a lack of oxygen, so too are the other parts of the brain. Brain stem death (BSD) has been accepted by the courts in the UK as the legal definition of death. This was first apparent in the case of *Re A* (1992) 3 Med.L.R. 303 involving a young boy taken to hospital with head injuries suggesting a non-accidental injury. He was placed on a ventilator, but was subsequently declared brain stem dead. His parents wanted him to be maintained on the ventilator to enable their own experts to examine him in the light of potential legal proceedings. The court refused, however, holding that the boy was legally dead and the doctors would not be acting unlawfully by disconnecting the ventilator. This judicial acceptance of BSD as legal death was subsequently confirmed by the House of Lords in *Airedale NHS Trust v Bland* [1993] 1 All ER 831, a case concerning a young man—Anthony Bland—who suffered severe brain damage in the Hillsborough disaster and was left in a persistent vegetative state (PVS). The Law Lords confirmed that this man was still legally alive. Lord Keith, for example, said that "In the eyes of the medical world and of the law a person is not clinically dead so long as the brain stem retains its function".[2]

It seems clear that BSD is accepted as the legal definition of death in the UK. Globally, a focus on the brain in defining death has broad acceptance, although some cultures, including Orthodox Jewish, Native American, and Japanese cultures reject it.[3] Typically, this rejection is founded upon a discomfort in regarding someone as dead if he or she is still breathing, whether artificially assisted or not. Perhaps for the same reason, brain stem death is not without its critics even in societies which have accepted the brain death concept, many of whom argue that it is counter-intuitive to classify an individual with a heartbeat as dead.[4] This view implies that the law currently regards a dying patient as already dead. If true, this would be a serious encroachment into any ethical or legal protection of life, such as through a right to life. However, as destruction of the brain stem is irreversible and the last part of the brain to be effected by a lack of oxygen, most, including the medical profession, judge BSD as a sign of death rather than dying. There are also some commentators who oppose BSD from the opposite perspective, arguing that we should go further and accept the concept of higher-brain death.

HIGHER-BRAIN DEATH: ITS PROPONENTS AND PROBLEMS

Proponents of higher-brain death argue that it is the irreversible loss of consciousness which signifies the end of life. An example of such a loss is a patient in persistent vegetative state (PVS). This condition entails irreversible damage to the higher brain when the brain stem is still functioning. The normal functioning of the brain stem means that the patient may be breathing independently, but the destruction of other parts of the brain means that the patient will have no awareness or consciousness of the world around him or her. It is the tragic condition suffered by Anthony Bland in the Hillsborough disaster and has challenged the courts many times since then. The PVS condition therefore poses considerable ethical and legal dilemmas across the world. PVS patients do not meet the criteria for either whole brain death or brain stem death but everything that made that patient a person has gone: memories, the ability to communicate, conscious awareness. Is this patient really still alive, or is death of the person that he or she used to be a sufficient criterion for the end of life?

Even in the *Bland* case mentioned above, some judges are uncomfortable with leaving the issue at an unambiguous acknowledgment of the patient being alive. Lord Goff, for example, raises a doubt about this conclusion when he states that the patient's condition "is such that it can be described as a living death".[5] He proceeds to outline the reason for the introduction of the BSD concept, explaining that "because, as a result of developments in modern medical technology, doctors no longer associate death exclusively with breathing and heartbeat". This, however, is a misleading diversion because in the traditional cardio-respiratory definition of death, Bland is also still alive because he is clearly still breathing. The "living death" concept introduced by Lord Goff seems to underlie many of the judgments and adds unnecessary ambiguity to the legal situation. Hoffmann LJ in the Court of Appeal fell into the same trap, confusingly stating that the patient's "body is alive but he has no life".[6] While these judicial comments raise concerns about the full extent of the law's protection for a human life in the "twilight zone" between life and death, the judges were not advocating a change in the legal definition of death. Some commentators do, however, adopt such an approach.

For such commentators, patients in PVS are already dead due to the loss of their higher brain functions, even though their body lives on.

Jeff McMahan, for example, argues that this organism should be treated as a dead body because:

> a mere organism does not have interests and cannot itself be benefited or harmed. To end its life is no more objectionable than it is to kill a plant, provided that what is done does not contravene the posthumous interests of, or manifest disrespect for, the person who once animated the organism.[7]

Such an approach makes a clear distinction between the "person" and the "organism" and, as such, is part of a broader ethical movement to distinguish between a person and a human being. The so-called "personhood theory" proposes that not all human beings are "persons" with rights. Although the exact requirements of personhood tend to vary between writers, they all focus on a disembodied mind. Consciousness is widely regarded as a minimum characteristic, and other proposed criteria include capacity for reason,[8] capacity to value one's own existence,[9] and moral agency.[10]

Personhood theory's focus only on a person with some degree of capacity takes the "Cartesian" model to its extreme manifestation. The seventeenth-century philosopher René Descartes explains reality as consisting of only *res extensa* (encompassing the corporeal body) and *res cogitans* (encompassing the mind). Not only are the body and mind thus distinct under Cartesian dualism, but the body is also subordinated to the mind, meaning that cognitive rationalisation dominates. Indeed, both Kant and Locke utilise a concept of a dominant mind over a mechanized body in order to establish a focus on rationalism. Unfortunately, the dominance of the rational mind over the emotional body under Cartesian dualism has gender-specific implications. As Shildrick notes, women are traditionally viewed as more intimately associated with their bodies and as "intrinsically unable to transcend them".[11] For example, hormones, PMT, pregnancy, menopause, "hysteria", and anorexia are just some of the ways in which a woman's body has, over the centuries, been regarded as affecting her rational mind. If a person is morally valuable because of the dominance of a rational mind over an unreliable body, women may face greater hurdles in maintaining and proving that distinction.

A further significant problem with the personhood theory is that it either includes other species within the concept of personhood (not necessarily objectionable in itself but requiring significant changes to our treatment of other species) or it excludes many human beings, including neonates and PVS patients. Proponents of the personhood theory, such as Peter Singer and John Harris, seem to be comfortable with excluding these categories of human beings from personhood, and thus from moral status. However, this does not sit easily with a human rights perspective. International human rights law places great value on underlying principles, such as equality of rights and respect for human dignity. Indeed, the underlying principle of human rights law is that all human beings are entitled to the same fundamental rights due to their status as human beings, regardless of distinctions such as nationality, race or gender. While very few human rights are absolute in their legal protection (and thus, for example, can be infringed where it is proportionate and necessary to do so), they do have universal application. The exclusion of a category of human beings from the protection of human rights law due to a particular physical or mental characteristic of those individuals is irreconcilable with equality of rights. Of course, if those individuals are no longer living human beings, they would not be entitled to equal protection with the living. The (legal) line between life and death is thus fundamental.

Death and the Right to Life

Every human being has a legally enforceable right to life. This right can be found, for example, in Article 2 of the European Convention of Human Rights (ECHR) and Article 6 of the International Covenant of Civil and Political Rights. Article 2 ECHR is also protected in domestic law by means of the Human Rights Act 1998. This does not, of course, mean that we can require our government to keep us alive indefinitely. It is an inherent fact of life that we will all die. Nonetheless, the right to life does impose a variety of obligations on the state. At its core, the right to life prohibits unjustified killing by the state. However, increasingly its interpretation requires far more than that core minimum, including positive obligations to take appropriate steps to safeguard the lives of those within its jurisdiction.

In *Osman v United Kingdom* (1998) 29 E.H.R.R. 245; Reps 1998-VIII at para 115, the European Court of Human Rights held that Article 2(1) "enjoins the state not only to refrain from the intentional and unlawful taking of life, but also to take appropriate steps to safeguard the lives of those within its jurisdiction". This means that state authorities must do all that could reasonably be expected of them to avoid a real and immediate risk to life of which they have, or ought to have, knowledge, although the Court did recognise that this obligation must be interpreted in a way which does not impose an impossible or disproportionate burden on the authorities.[12] A government committed to the right to life must, therefore, not merely refrain from killing, but also govern in a manner that seeks to preserve human life wherever reasonably possible. Arguably, the most important principle underlying the right to life is not the sanctity of life but rather a requirement of respect for all human life.[13]

The right to life protects all living human beings (with some lingering ambiguity about its application before birth (*Vo v France* (App. 53924/00), 8 July 2004 [GC], (2005) 40 EHRR 259, ECHR 2004-VIII). As discussed above, under the BSD definition of death, that includes patients in PVS whose brain stem continues to function even if the aspects of their higher brains which made them who they were have been destroyed. Does the state have an obligation under the right to life to maintain the life of patients in PVS? It would appear not. The landmark English case of *Airedale NHS Trust v Bland* confirmed that it is lawful to withdraw life-sustaining treatment (commonly artificial nutrition and hydration (ANH)) from a patient in PVS. The key to the legality of such a course of action is that it is no longer regarded as in the best interests of a patient in PVS to receive the life-sustaining treatment. In the absence of patient consent, medical treatment can only be provided if it is in the best interests of the patient. The best interests test was developed at common law in a very different context (non-therapeutic sterilisations) but is now at the core of the statutory regulation of persons who lack capacity. Section 4 of the Mental Capacity Act 2005 requires that all decisions about persons lacking capacity are taken in their best interests and includes a checklist of factors to be taken into account in determining that. When applied to a patient in PVS, this means that a decision has to be taken whether continued treatment (usually comprising ANH) is in the patient's best interests and, if it is not, then it must (rather than may) be withdrawn. Under the Mental Capacity Act, the best interests test is more patient-focused than before, with a requirement to take into account the patient's own past wishes and feelings, values and beliefs, rather than merely medical evidence as to prognosis. The latter is still likely to be very significant, however, and thus it is still not an entirely subjective test of interests. In *Bland*, the Law Lords were adamant that the relevant question was whether treatment is in the patient's best interests rather than whether death is in the patient's best interests but, given that a withdrawal of life-sustaining treatment will inevitably cause death, this is a somewhat meaningless distinction.

The *Bland* case was decided before the enactment of the Human Rights Act 1998 (HRA) and thus the court gave no weight to issues of patient rights, including the right to life. However, subsequent to the HRA's coming into force, the courts have sought to reconcile the *Bland* judgment with the right to life. In *NHS Trust A v M; NHS Trust B v H* [2001] 1 All ER 801, Butler-Sloss LJ explained that there is no violation of Article 2 ECHR's right to life when life-sustaining treatment is withdrawn because it is no longer in the patient's best interests. This approach to interpretation of Article 2 seems to have been plucked from thin air given that there is no precedent for reading the state's obligations under Article 2 as subject to a best interest determination. In terms of the right to life, this may look suspiciously like we have a right to life until it is not in our best interests to continue living. While this may be a sensible, even justifiable, approach in the context of PVS patients, it is surely worrying in more general terms? Who is to decide when our lives are no longer in our best interests? And how will they know?

The withdrawal of medical treatment leading to death needs to be acknowledged as a legitimate exception to the right to life, but it is a pity that more explicit reasoning has not yet been provided to reconcile the right to life with end-of-life decision-making. As the state is only ever obliged to take reasonable steps to preserve life, it may be that a focus upon both autonomy and quality of life could cast a clearer light upon when it is no longer reasonable to preserve a life. Even in the context of a PVS patient, however, the right to life and other human rights do still have meaning and value. A doctor might be able to withdraw treatment to allow the patient to die, but it would not be lawful for a hospital intruder to shoot the patient in the head, nor for the hospital to throw the patient out onto the street. We would not accept a degrading use of the patient's body even though he is not aware of it. Thus, we do still value this human being—and the law does too. It is a living person, not just an empty shell, or an organism which the "person" has vacated.

Life, Death, and Embodied Selves

The most objectionable element of the personhood theory's approach, discussed above, is that it does not regard patients in PVS (or indeed many other living human beings) as persons with moral value.

Personhood theory entirely overlooks the value of the human body in its rush to remove rights from PVS patients. By contrast, embodiment theory focuses upon the whole person as a union of body and mind.[14] It recognises the interaction and relationship between our mental selves and our bodies, and gives value to the body in contrast with the traditional division of body and mind under Cartesian dualism, which views the body as little more than a machine. A focus on the embodied self would put the body back into the equation and, in relation to the definition of death, would rule out any move towards higher brain death.

As noted above, higher-brain death proponents view a living body devoid of the mind as already deceased. This is in stark contrast to an embodiment approach which views both body and mind as crucial to moral status and legal respect. As Martin Pernick explains, the "controversy between advocates of whole-brain and higher-brain criteria for diagnosing brain death often reflected a much older conceptual contest over whether mental activity or bodily integration constituted the essence of human life".[15] Advocates of higher-brain death recognise only the human mind and not the human body as being morally valuable. As such, this definition of death would rest upon a concept of the disembodied self. It connects naturally with personhood theory which similarly affords respect to the mind as distinct from the body. Indeed, as Harris confirms, under the personhood theory, persons need not be organic life forms at all.[16]

So, what of an organic life form devoid of a functioning mind? Is it conceivable that a deceased person residing in a living body could be buried or cremated? This is the inevitable consequence of both personhood and higher-brain death theories. Lizza explains why it is not an insurmountable hurdle for him as he regards the continuation of the organism that once constituted the person as a non-critical issue, arguing that mere organic integration is insufficient for the continued life of a person.[17] By distinguishing between the life of the person and the life of the human organism, Lizza is able to envisage the burial of a living and breathing human body: "Instead of a person's death resulting in remains in the form of an inanimate corpse, a person's remains can now take the form of a living being devoid of the capacity for consciousness and any other mental function".[18] Such an approach has so far deviated from legal concepts of human rights and human dignity as to be irreconcilable with existing international and national law commitments to human rights law, as well as ethical and moral obligations to humankind. The law unsurprisingly rejects such an artificial distinction between deceased persons and living human bodies and instead, as explained above, regards the irreversible death of the brain (or brain stem) as the point of death. At this stage, and not before, legal obligations to respect the rights of a human being cease. Thus the right to life, and the (albeit limited) protection it affords, ends only when life itself has ended.

The question of why human life is protected in the first place—why human life matters—seems to be closely tied to the concept of consciousness. The higher level of consciousness enjoyed by humans as compared to many other species is, arguably, the underlying reason why human life should be regarded as more valuable and given greater protection, morally and legally, than the life of a virus or a plant or an animal.[19] There is not, as yet, any clear explanation for how consciousness arises or why, although there is an indisputable link with electrical brain activity. As Merlin Donald explains, "conscious effort is the single most reliable predictor of the patterns of brain activity".[20] However, science still cannot tell us why this is so: "Brains that pulse with certain patterns of electrical activity are conscious. Why? They just are".[21] Human life matters, it is argued, because of human consciousness.[22] But this does not inevitably lead to the personhood theory's controversial approach of excluding certain human beings from moral status and legal protection due to their loss of a characteristic such as consciousness. If human life matters, then it *always matters*, regardless of personhood, rationality, moral agency, capacity or consciousness. The life of an individual human being matters not because that organism is sentient or rational (or free of pain, or values its own existence) but because it is a human life.[23] This point is supported by the ethical and legal principle of equality which is well established in the field of human rights. A core requirement for an individual human being to be regarded as possessing a life may be regarded as basic integrative functioning of the organism. In other words, (from viability) until brain death, while a human organism has the potential to function in an integrative manner, an individual has a life equal to that of all other human organisms. In terms of defining death, this means that life ends only when the human organism—the body and mind together—dies. This cannot sensibly require the death of all of the body's cells, but rather the death of the organism as a whole. In other words, life comes to an end when the integrative action between the organs of the body is irreversibly lost. The death of the brain, or the brain stem, is one, and perhaps currently the best, means of discerning that end. It is fitting, therefore, that the law has adopted this stage for the legal definition of death.

CONCLUSION

The law has a crucial role in regulating issues of life and death. In addition to its essential role in clarifying legal responsibilities, it is an important influence on social and ethical perspectives regarding the inevitable endpoint of life. Furthermore, the entire concept of human rights law is based around the idea that all human beings are entitled to equal legal protection for a range of rights and freedoms, thus necessitating an unambiguous dividing line between a rights-holder and a deceased body. Death is not unambiguous, however, at least not in its appearance to modern medical technology and understanding of the human mind and body. The point at which the law draws the line between life and death, or more specifically between dying and death, is always likely to be controversial. The contemporary focus on the irreversible destruction of the brain—brain stem death in the UK—builds upon the current state of medical technology and its ability to revive certain parts of the human body. It also fits well with an embodiment approach to valuing human life which strives to include both mind and body within conceptions of the person and moral status. The alternative, albeit increasingly influential, personhood theory, with its singular emphasis upon the mind, would lend support to a different conception of death: one that hinges upon the destruction of the higher-brain and certain mental capacities. As tragic as such a transformation in life can be, it is not appropriately regarded as a death of a human being, and to label it a death of a person is dangerously misleading. We are not just our minds, but also our bodies, which serve as our homes, our transport, our clothing, our identities.[24] They are also inevitably our ultimate cause of death, for we will not survive the loss of our bodies. This focus on the embodied self does not necessitate a striving to sustain all permanently comatose human lives. The withdrawal of life-sustaining treatment is not only lawful under specific circumstances, but is also sometimes ethically appropriate, morally good, and respectful of the human being's rights. But let us never forget, or worse ignore, that the human being who has lost so much of what made her an individual, is still alive and entitled to a right to life; a right that is always limited, both in terms of state obligations and its application to mortal beings.

NOTES

1. "A Definition of Irreversible Coma: Report of the Ad Hoc Committee of the Harvard Medical School to Examine the Definition of Brain Death", *Journal of American Medical Association*, 205 (1968), p. 337.
2. *Airedale NHS Trust v Bland* [1993] 1 All ER 831, p. 859.
3. See David Novak, *The Sanctity of Human Life* (Washington, D.C., 2007); Noel Williams, *The Right to Life in Japan* (London, 1997); Clive Lawton, "Judaism". In Clive Lawton and Peggy Morgan, *Ethical Issues in Six Religious Traditions* (Edinburgh, 2007).
4. See Martyn Evans, "Against the Definition of Brainstem Death". In Robert Lee and Derek Morgan eds., *Death Rites: Law and Ethics at the End of Life* (London, 1994), pp. 1–10.
5. *Bland*, p. 865.
6. Ibid., p. 850.
7. Jeff McMahan, "Brain Death, Cortical Death and Persistent Vegetative State". In Helga Kuhse and Peter Singer eds., *A Companion to Bioethics* (Oxford, 1998), p. 258.
8. See Peter Singer, *Rethinking Life and Death: The Collapse of Our Traditional Ethics* (Oxford, 1995).
9. See John Harris, *The Value of Life: An Introduction to Medical Ethics* (London, 1985).
10. See Immanuel Kant, *Groundwork of the Metaphysics of Morals*, Lawrence Pasternack ed., (London, 2002).
11. Margrit Shildrick, *Leaky Bodies and Boundaries: Feminism, Postmodernism and (Bio) Ethics* (London, 1994), p. 26.
12. *Osman v United Kingdom* (1998) 29 E.H.R.R. 245, para 116.
13. Elizabeth Wicks, "The Meaning of Life: Dignity and the Right to Life in International Human Rights Treaties", *Human Rights Law Review*, 12 (2012), p.199.
14. Alistair V. Campbell, *The Body in Bioethics* (London, 2009), p. 4.
15. Martin S. Pernick, "Brain Death in a Cultural Context: The Reconstruction of Death 1967-1981". In Stuart J. Youngner, Robert M. Arnold, and Renie Schapiro eds., *The Definition of Death: Contemporary Controversies* (Baltimore, 1999), p. 12.
16. John Harris, "The Right to Die Lives! There is No Personhood Paradox", *Medical Law Review*, 13 (2005), p. 389.
17. John P. Lizza, *Persons, Humanity and the Definition of Death* (Baltimore, 2006), p. 13.
18. Ibid., p. 15.
19. See Elizabeth Wicks, *The Right to Life and Conflicting Interests* (Oxford, 2010).

20. Merlin Donald, *A Mind So Rare: The Evolution of Human Consciousness* (New York, 2001), p. 178.
21. Ibid.
22. See Wicks, *The Right to Life*.
23. Ibid., p. 16.
24. See Elizabeth Wicks, *The State and the Body: Legal Regulation of Bodily Autonomy* (London, 2016).

Author Biography

Elizabeth Wicks is a Professor of Human Rights Law in the Leicester Law School at the University of Leicester. Her main research interest is in human rights in healthcare. Her latest book, *The State and the Body: Legal Regulation of Bodily Autonomy*, was published by Hart in 2016.

Open Access This chapter is licensed under the terms of the Creative Commons Attribution 4.0 International License (http://creativecommons.org/licenses/by/4.0/), which permits use, sharing, adaptation, distribution and reproduction in any medium or format, as long as you give appropriate credit to the original author(s) and the source, provide a link to the Creative Commons license and indicate if changes were made.

The images or other third party material in this chapter are included in the chapter's Creative Commons license, unless indicated otherwise in a credit line to the material. If material is not included in the chapter's Creative Commons license and your intended use is not permitted by statutory regulation or exceeds the permitted use, you will need to obtain permission directly from the copyright holder.

CHAPTER 9

The Last Moment

Jonathan Rée

For many years, studies of death have been dominated by the distinction made by Philippe Ariès between modern ways of dying, hidden away in a hospital room, and traditional death-bed rituals, where friends gather round and celebrate the process with stories, music, art and prayer.[1] The difference is obviously connected with a re-orientation in metaphysical opinion about the afterlife: if you believe that the dying person is going to live on and meet you again, then you will do your best to part on good terms; but you have no reason to bother if you are a materialistic modernist who thinks that death means total annihilation.

The contrast is somewhat overdrawn, however, not only historically (as we all know) but metaphysically too. If you have a robust belief in individual survival, you will still be prey to doubts about life on the other side, and once you try to envisage it in detail—what people will look like, whether they will change over time, and how friendships will be conducted—it is liable to lose both its attraction and its plausibility. On the other hand, if you are a secular rationalist grieving for someone you love, your unconscious can be counted on to supply you with fantasies about escapes from death and renewed encounters. If we met someone whose

J. Rée (✉)
London, UK
e-mail: jonathan.ree@gmail.com

attitude to death was untouched by conflict, doubt, and an uneasy sense of mystery we would, I think, conclude that they were not really human.

The scenarios of post-mortem encounter that play themselves out in our imaginations are of enormous interest, both historically and psychologically; but they tend to distract us from a phenomenon that is perhaps equally significant, and equally fundamental: from what might be called the *fascination of the last moment*. When we get news of the death of strangers—famous people, perhaps, or victims of a mass-shooting or a plane crash—we find it hard not to wonder what it was like for them in their last hour, their last minute, or their last second: what were they thinking about; what did they know; and how did they feel? In the case of close friends and family, we will speculate rather more intrusively: were they inwardly angry or were they serene; were they in pain; and as they departed this world, did they cast a glance back at someone in particular—for example, did they spare a thought for me?

If we are modern and rational, we will try not to be so sentimental. Why should any special interest attach to the last moment of a life? What makes it different from any other segment of time? All of us are subject to ups and downs in our moods, and why worry about which part of the cycle we are in when it stops? The historians among us may offer explanations in terms of inherited religious traditions which portray us facing our maker and being made to answer for how we spent our life on earth. In that case, our dying state of mind might possibly swing the case for us: a lifetime of wickedness could be cancelled by last-minute repentance, and perhaps the converse holds as well. The absurd disproportion between the dying moment and eternal punishments, or rewards, may have been a challenge for subtle theologians, but the deathbed industry got on with the business of conversion, confession, indulgence, unction, absolution and prayer. It is easy to make fun of these last-chance rituals, comparing them perhaps to stand-and-deliver academic exams, or a game of roulette; but I would like to suggest that the fascination of the last moment has roots that reach far deeper than the contingencies of religious doctrine.

Death has a special presence in everyone's life, even for those who encounter it as an everyday reality—priests, doctors, nurses, and undertakers, or, in a different way, historians and archaeologists. If we are living like human beings as well as fulfilling our occupational roles, then every moment of our existence has the characteristic that Martin Heidegger called *Sein zum Tode*, or being-towards death.[2] Our activities get their significance from our sense of them as episodes in a life-story that began earlier than we can remember and will end with our death even if we never think about it explicitly; death will always be, as a phenomenologist would put it, the "horizon" of our existence. Hence, the peculiar thrill of witnessing a death, or of death as a spectacle; however much we differ from each other in particular ways, the anticipation of dying is something we all have non-contingently in common. We can hardly stop ourselves feeling some kind of sympathy, or personal involvement, with people on the brink of death; we identify with them, because we know that our own turn will come one day.

***

This perspective—call it existential or phenomenological if you like—may cast some light on the salience of the moment of death in fine art, high tragedy, and grand opera, or more particularly, in narrative fiction, since the middle of the nineteenth century. Take Leo Tolstoy's 1886 story, "The Death of Ivan Ilych", which opens with a character called Peter Ivanovich learning that his old friend Ivan has died, still young, though no longer full of promise. Even though Peter Ivanovich has known Ivan Ilych all his life—they used to play together as little boys—he receives the news with apparent indifference. He pays an obligatory visit to Ivan's widow, even though he dislikes her; but to his surprise he is terribly moved, at least briefly, when she tells him how Ivan passed his last hours: he "screamed incessantly", she tells him, and was "conscious all that time", right up to "the last moment".

> 'Three days of frightful suffering and then death! Why that might suddenly, at any time, happen to me', he thought, and for a moment felt terrified. But...the customary reflection at once occurred to him that this had happened to Ivan Ilych and not to him...after which reflection Peter Ivanovich felt reassured...as though death was an accident natural to Ivan Ilych but certainly not to himself.

But the act of repression requires an effort that Peter Ivanovich is unable to sustain; despite his outward nonchalance, he knows that death is the great leveller, eventually making equals of us all.

Ivan Ilych had known it too; the truth had impressed itself on him as he talked to the peasant lad who looked after him in his final illness. The boy seemed happy in his work, however repulsive his duties were, and he explained that he was sustained by the knowledge that "we shall all of us die", telling Ivan that he "did not think his work burdensome… because he…hoped someone would do the same for him when his time came".

But all of us know—we the readers, and Ivan, Peter, and the peasant boy—that the fellowship of death goes deeper than some intergenerational compact about end-of-life care. Ivan died in terrible pain, but his spiritual anguish was far worse: "I am leaving this life", he said to himself, "with the consciousness that I have lost all that was given to me and it is impossible to rectify it". He "had not spent his life as he should have done", and he "struggled", we are told, "as a man condemned to death struggles at the hands of the executioner, knowing that he cannot save himself".[3]

The analogy between being condemned to death in the special sense of facing judicial execution, and being condemned to death in the way all of us are, has tremendous emotional and literary resonance. Dr Johnson was right, no doubt, to say that "when a man knows he is to be hanged in a fortnight, it concentrates his mind wonderfully".[4] But that tells only half the story, because the prisoner will not be the only one counting down the minutes that remain: anyone who knows about the impending event will be doing the same. Crowds of us will gather at the spot, in imagination if not reality, to project ourselves into the consciousness of the prisoner and rehearse for the time when our own end is near.

Claude Lanzmann (the filmmaker who made the documentary film *Shoah* (1985)), began his autobiography by saying that he could not remember a time when he was not entranced by the idea of being sentenced to death. In 1938, when he was 13, he read about the death by guillotine of a murderer called Eugen Weidmann in the street outside the prison in Versailles. (It would be the last public execution in France, and the newsreel is available online). Lanzmann was struck by the resemblance, such as it was, between his own name and that of the man about to be decapitated, and he was never able to shake off the fantasy that a similar fate awaited him; and as he wrote his memoirs seventy years later, he still imagined himself as a prisoner on death row, struggling to give an account of himself before his appointment with death.[5]

But it is not just Lanzmann; law has always aspired to the condition of theatre, and judicial executions are its masterpiece, generating an appetite for tales of the prisoner's last moments. But *execution narratives*, as they are sometimes called, come in two very different kinds. Some are *hostile*: they work to place those condemned to death at a great distance from us, beyond the pale of common humanity. Take, for example, a pamphlet about the execution in 1606 of the conspirators in the gunpowder plot—a "horrible and abominable Treason", as we are informed: "detestable in the sight both of God and man" and "odious in the eares of all humane Creatures". We are invited to shudder at the thought of their "bewitched hearts", to wonder at their refusal to exercise the "true repentance, that in true Christians may be required" and, in short, to marvel "that so many monsters in nature, shoulde carry the shapes of men".[6]

Then there are the *sympathetic* execution narratives, which ask their readers to identify with those about to die, and enter into their inner world; the corpus could be thought of as reaching back to Plato on the death of Socrates, or to the gospel accounts of the crucifixion, and it would include hagiographies of Christian martyrs, and heroic accounts of royal beheadings. The genre seems to have remained distinctly aristocratic in focus until the middle of the nineteenth century, when it took a turn towards literary realism, and its protagonists became more plebeian.

The closing pages of Charles Dickens's *Tale of Two Cities* (1859) provide a transitional case, with Sydney Carton achieving greatness by choosing to die in the place of Charles Darnay (perhaps in the place of us all) and eliciting our tears (for him, for us) with the supreme pathos of his inner monologue ("It is a far, far better thing that I do, than I have ever done; it is a far, far better rest that I go to than I have ever known") on the scaffold. The situation poses a problem from the point of view of narrative technique, and the solution chosen by Dickens was bold if not crude: if Sydney had "given utterance" to his thoughts, he wrote, "they would have been these".[7] The problem seems to have defeated Herman Melville, who was never able to complete his complex and multi-perspectival execution narrative, *Billy Budd*. His contemporary Ambrose Bierce was apparently unfazed by the difficulty, and his "An Occurrence at Owl Creek Bridge" (1890), became the founding classic of sympathetic execution narratives. Without fussing over the question of narrative point of view, Bierce offered direct access to the consciousness of a "man who was engaged in being hanged" and the lifelike world he entered as he "closed his eyes in order to fix his last thoughts on his wife and children".[8]

Albert Camus offered a fully first-person execution narrative in the last chapter of *L'Étranger* (1942), where the unrepentant murderer Meursault fills his notebook with reflections on the prospect of being guillotined the next morning. "I regret not paying more attention to tales of execution", he says, as his imagination scurries around searching for a way out; for example, if the blade of the guillotine gets jammed in mid-fall. He also remembers that his father once attended a public execution and ended up sick with rage and fear. "From that point on my father rather disgusted me", he recalls, but now he thinks he understands:

> How could I have failed to see that there was nothing more important than an execution and that in fact it was the only really interesting thing in a human life? If I got out I would certainly make it my business to go and watch every public execution I could.

But he knows it is not going to happen; instead of witnessing the execution of others, he will have to participate in his own, and he ends his account with the words: "I only hope there will be plenty of spectators at my execution, and that they will greet me with cries of hatred".[9]

Grammatically, *L'Étranger* is an exercise in first-person narration, but it keeps spinning round to a third-person perspective in a way that seems to be characteristic of sympathetic execution narratives. Or equally the other way around: a third-person narration will directly evoke a first-person perspective—as for example in George Orwell's essay "A Hanging", which describes an incident in Burma in the 1920s. As an officer of the British Imperial Police, Orwell was required to supervise the execution of a poor Hindu, after which he met up with his British colleagues and found himself laughing far too much as he tried to conceal his anguish at what he had seen and done:

> It was about forty yards to the gallows. I watched the bare brown back of the prisoner marching in front of me…At each step his muscles slid neatly into place…and once, in spite of the men who gripped him by each shoulder, he stepped lightly aside to avoid a puddle in the path.

That little gesture, the deft avoidance of the puddle, stuck in Orwell's memory like a dart:

> It is curious, but till that moment I had never realized what it means to destroy a healthy, conscious man. When I saw the prisoner step aside to avoid the puddle I saw the mystery, the unspeakable wrongness, of cutting a life short when it is in full tide. This man was not dying, he was alive just

as we are alive...His eyes saw the yellow gravel and the grey walls, and his brain still remembered, foresaw, reasoned—even about puddles. He and we were a party of men walking together, seeing, hearing, feeling, understanding the same world; and in two minutes, with a sudden snap, one of us would be gone—one mind less, one world less.

Orwell flips the third person into the first ("one of us would be gone") and we are inside the world of the condemned man.[10]

The shuttle between different points of view at the execution scene becomes more elaborate in the final chapter of Truman Capote's *In Cold Blood* (1966), which tells the story of the gruesome murder of a farming family—Herb Clutter and his wife and two children— in Kansas in November 1959, and of the two obtuse young lads, Richard Hickock and Perry Smith, who committed the crime in the mistaken belief that they would find large sums of money in the house. When their trial ends and Hickock and Smith receive their death sentences, they laugh loudly as they are taken to their neighbouring prison cells, out of sight of each other, but not out of earshot. After five years of incarceration, their appeals are exhausted, and Capote describes how they spent their last day—Tuesday 13 April 1965—looking forward to being hanged one after the other, in alphabetical order (as they had chosen), just after midnight. They both ordered a large meal: shrimps, fries, and strawberries and cream. When the time came, Smith was, as usual, rather withdrawn, but Hickock, who was the intellectual of the two, extended a warm welcome to those who came to witness his death ("nice to see you!"), and expressed disappointment that no members of the Clutter family had bothered to come ("as though he thought the protocol...was not being properly observed"). "I just want to say I hold no hard feelings", he said as he stepped up to the gallows: "you people are sending me to a better world". Then it was Smith's turn: "It would be meaningless to apologise for what I did", he said: "but I do: I apologise". And then, as Capote puts it, "the thud-snap that announces a rope-broken neck".[11]

***

The rise of the sympathetic execution narrative went alongside a striking alteration in ordinary attitudes to the death penalty, and was, no doubt, in part responsible for it: a public that has learnt to identify with the prisoner as the last seconds tick away is likely to find the very idea of judicial execution unconscionable and to lose any capacity to imagine how enlightened opinion could ever have taken a different view. John Stuart Mill was acutely aware of swimming against the tide when, during his brief career as a member of parliament, he gave a magnificent speech on the subject in 1868. He was embarrassed, as he put it in his *Autobiography*, to find himself advocating a position "opposed to what then was, and probably still is, regarded as the advanced Liberal opinion". It struck him as inconsistent to say that every citizen has a right to life, which the state can never abridge; after all, every citizen has a right to liberty too, but no one has any problem allowing the state to withdraw it in the case of certain kinds of crime, and for that matter extreme insanity. He therefore defends the death penalty (provided it is restricted to "atrocious crimes" that have been proved beyond all possible doubt), and he defends it on enlightened, modern grounds—"on the very ground on which it is commonly attacked—on that of humanity to the criminal". We may think it would be merciful to spare the life of the criminal; but, Mill asks, "what kind of a mercy is this?"

> There is not, I should think, any human infliction which makes an impression on the imagination so entirely out of proportion to its real severity as the punishment of death...Is death, then, the greatest of all earthly ills?...Is it, indeed, so dreadful a thing to die? Has it not been from of old one chief part of a manly education to despise death—teaching us to account it, if an evil at all, by no means high in the list of evils....The human capacity for suffering is what we should cause to be respected, not the mere capacity of existing.

Mill's argument strikes me as impeccable: executing a murderer might indeed be an act of mercy, and some execution narratives could be cited in support of it. Camus's hero was, you might say, begging to be executed, and he would have been horrified to be reprieved; and as for Hickock and Smith, I, for one, find it hard to put down Capote's book without thinking that it would have been cruel to condemn these two young men, both in their thirties, to "existing" for its own sake; however long they might have lived, they could never have amounted to anything more—in their own eyes or those of others—than exceptionally stupid and cruel murderers. If they had any sense, they would have wanted to take their own lives, and the death sentence was enabling them to get what they wanted. But then, as Mill admits, the question turns, for many of us, not on the "real severity" of the punishment, but on the way in which it "makes an impression on the imagination".[12]

***

I conclude with another sympathetic execution narrative—a description of a hanging in Georgia, USA, in 1893, by the French writer Paul Bourget. The prisoner was Henry Seymour, a young "mulatto" (as Bourget calls him) who had once enjoyed the affection and patronage of a gentleman called Colonel Scott, accompanying him on some of his hunting expeditions. But Seymour had gone off the rails, first committing a murder for which he was condemned to death, and then killing one of his warders and escaping. He was apprehended by Scott himself, who gently dressed his wounds and returned him to prison, leaving him with a bottle of fine whiskey—the same brand they used to drink together when they went hunting—urging him to finish it before he dies. The encounter with the hangman was fixed for the following day, but with a choice of times—any time between 9am and 4pm—and Seymour settles for 1:45pm, so that he can enjoy a good meal, waited on by the sheriff in person.

Bourget was issued with a ticket to witness the execution, but as a man of sensibility, he screwed it up and threw it away in the street. But then he reflected on the seriousness of what was about to be enacted in the prison:

> Any person of culture who has entertained the thought of observing a public execution has probably been through the same nervous emotions as I. The mysteries of death, of moral responsibility and of justice and social right...together with the most intimate frisson of existence, as at the approach of inexorable tragedy...are all wrapped up in an execution of this kind.

In the end he retrieves his ticket, and on admission to the prison, he is able to gaze on Seymour—a lean figure of arresting beauty, reminiscent of a "bronze statue", except for his warm vitality and the "simple play of his muscles"—and he observes him tucking into a plateful of fried fish. Seymour notices Bourget and says with a smile: "I will carry with me a belly full of fish, where I go!", before washing his hands, combing his hair, and walking out to greet the witnesses gathered round the gallows. (His wife and young children were not admitted, but they were waiting in the street outside.) He mounts the scaffold with a firm tread, and if he feels any anxiety he shows it only for a moment, when the cigar he kept for the occasion falls from his lips; and once the noose is round his neck, he turns to Bourget, Scott and the rest, utters a brief prayer, and finishes with "I am all right now", and "good bye everybody...good bye colonel".[13]

Bourget was impressed by the "irony" of the scene: by the fact that an ignorant mulatto who seemed to live only for whiskey and fried fish had faced his executioner with a display of courage that was "suddenly ennobled by a touch of the ideal".

> What an irony, that a man of this character—an orang-outang with the capacity to...speak—suddenly achieved what philosophers regard as the supreme fruit of their teaching: resignation in the face of the inevitable.

Philippe Ariès refers to Bourget's story and finds his analysis absurd: Bourget failed to appreciate that Seymour did not belong to the same modern rational world as him, and that his behaviour was simply a manifestation of the pre-modern tradition of "immemorial resignation in the face of death".[14]

This dispute strikes me as artificial. Seymour's death was part of a public ritual—a piece of legal and political theatre belonging to a particular place and time, and as such, it is open to investigation by the methods of the historian. But between the lines of Bourget's evolutionary racism, we can glimpse something more primordial: Seymour as someone just like us, trying to make sense of a difficult situation, and facing a death like any other, in which everyone can see a prefiguration of their own. He was not appropriating a few lines from an incongruous modernity, as Bourget seems to have supposed, but neither was he confined, as Ariès suggested, within the limits of pre-modern tradition. He had never really been traditional, and he was never going to be modern, but he was caught up, like everyone else, in the existential fascination of the last moment.

Notes

1. Philippe Ariès, *L'Homme Devant la Mort* (Paris, 1977), pp. 13–96.
2. Martin Heidegger, *Being and Time: A Translation of "Sein und Zeit"*, Joan Stambaugh trans. (New York, 1996), p. 216.

3. Leo Tolstoy, *The Death of Ivan Ilych: and Other Stories*, Aylmer Maud trans. (New York and London, 1960), pp. 95–156, 101–102, 138, 152, 154.
4. James Bowell, *The Life of Samuel Johnson*, Vol. 2 (London, 1791), p. 152.
5. Claude Lanzmann, *Le Lièvre de Patagonie* (Paris, 2009), pp. 17–19.
6. Cited in Leigh Yetter ed., *Public Execution in England, 1573–1868*, Vol. 1 (London, 2009), pp. 47–72, 48–49. See also Katherine Royer, *The English Execution Narrative, 1200–1700* (London, 2014).
7. Charles Dickens, *A Tale of Two Cities* (New York, 1910), pp. 376, 377.
8. Ambrose Bierce, *Terror by Night: Classic Ghost & Horror Stories* (Ware, Hertfordshire, 2006), p. 2.
9. Albert Camus, *L'Etranger* (Paris, 1942), pp. 165, 168, 186.
10. Cited in Joshua R. Farris and Charles Taliaferro eds., *The Ashgate Research Companion to Theological Anthropology* (London and New York, 2015), p. 320. In Gabriel Garcia Marquez, *Chronicle of a Death Foretold*, Gregory Rabassa trans. (London, 2014), p. 122, the author plays an extended game with the idea of an impending execution. In it, the dying Santiago, with his viscera tumbling out of his stomach, "even took care to brush off the dirt that was stuck to his guts".
11. Truman Capote, *In Cold Blood* (London, 2000), pp. 300, 331, 333.
12. John Stuart Mill, *Collected Works*, Vol. 1, John M. Robson and Jack Stillinger eds., (Toronto, 1981), p. 275; Vol. 28, John M. Robson and Bruce L. Kinzer eds., (Toronto, 1988), pp. 266–272.
13. Paul Bourget, *Outre-Mer (Notes sur l'Amérique)*, Vol. 2 (Paris, 1895), pp. 241–261.
14. Ariès, *L'Homme*, p.35.

Author Biography

Jonathan Rée is a writer and teacher who seeks to combine the methods of philosophy with those of history and literary theory. His works include *Proletarian Philosophers* (1984), *Philosophical Tales* (1987), and *I See a Voice* (1999).

Open Access This chapter is licensed under the terms of the Creative Commons Attribution 4.0 International License (http://creativecommons.org/licenses/by/4.0/), which permits use, sharing, adaptation, distribution and reproduction in any medium or format, as long as you give appropriate credit to the original author(s) and the source, provide a link to the Creative Commons license and indicate if changes were made.

The images or other third party material in this chapter are included in the chapter's Creative Commons license, unless indicated otherwise in a credit line to the material. If material is not included in the chapter's Creative Commons license and your intended use is not permitted by statutory regulation or exceeds the permitted use, you will need to obtain permission directly from the copyright holder.

CHAPTER 10

Afterword

Thomas W. Laqueur

King Lear gets it right for the ages:

> And thou no breath at all? Oh, thou'lt come no more,
>
> Never, never, never, never, never

he says holding the body of the dead Cordelia. Breath—or more precisely the possibility of future breaths—is still what matters, as it always has, in answer to the question "when is death?"

Before the widespread use of positive pressure ventilators in the last three or four decades of the twentieth century, making this sort of prognosis was, in general, not very hard. If there was no breath for a few minutes there would never—never, never—be breath in that body again. (There are only three "nevers" in the quarto text of King Lear; this suffices).

Of course, there were exceptional circumstances, increasingly recognised in the late eighteenth century, that might make it more difficult to say for sure that breath, after it had stopped, might not return and therefore to be sure that the condition of death—being dead—might not have

T.W. Laqueur (✉)
University of California, Berkeley, California, USA
e-mail: tlaqueur@berkeley.edu

been misdiagnosed. "When is death?" "Not now, not in this instance", the answer might be. People who had drowned and whose bodies were chilled could be "revived" [Anglo-Norman; Middle French; Latin: made to live again, to be given fresh life] or resuscitated—revived from a moribund state. Their breath could be restored.

The same might be the case for those who had succumbed to carbon dioxide—fixed air, or carbonic acid gas as it was called back then—that had replaced oxygen in ships' holds, fermenting vats, or other closed spaces. So too people who had stopped breathing for a few other select reasons: being struck by lightning, for example. "[T]he unfortunate objects are too often deserted, when they might every now and then be resuscitated by the various means employed", we read in an issue of the late eighteenth century *Transactions of the Royal Humane Society*, one of scores of organisations founded to teach the ways in which breath could be restored and a mistaken prognosis set aside. But even under such special circumstances, the window between the temporary cessation of breath—that is life—and its eternal absence was narrow—measured in minutes not hours.

Other instances besides these in the general category of "apparent death" seemed to present further important difficulties for deciding whether breath was truly gone forever, or just in temporary abeyance. Comas, for example, might mimic a deep sleep in which breath, presumably still present, might be so shallow as to be difficult to detect. Thomas Willis, eponymous discoverer of the anastomotic system, the circle of arteries that sits at the base of the brain, recognised this state in the late seventeenth century. Between the middle of the eighteenth and end of the nineteenth century, meanwhile, there was, in the western world, a minor epidemic fear of being buried alive. It was fed by gothic fiction, medical papers reporting on strange cases of mistaken death diagnoses, the popular press, and, as Brian Parsons suggests, the commercial interests of those selling remedies—coffins with bells that could be rung from inside and heard above ground.

But in fact, in cases of suspected apparent death, it would have been clear relatively quickly whether breath would "never, never, never" come again. The dead body, that is, the eternally breathless body, cools to the ambient temperature at a rate of about 1.5 °C per hour. It would not take long for it to feel cold. In very hot climates, rapid onset of decomposition would give the story away. Rigor mortis of the small muscles of the face starts within hours. The blood pools and things begin to fall apart. In short, a body whose breath is never to return comes to look dead because it is dead within a very short amount of time. Only in the modern age, when death certificates demand a precise hour of death is there a need for more precision.

In the 1960s a new, adjectivally limited, kind of death made its appearance—brain death—and with it, a new kind of answer to the "when" question. Lear's answer, resonant since the beginning of human contemplation on the subject—never, never, never, never, never more breath—seems to have been replaced by a novel, technologically and philosophically, mediated one that to lay people can seem strangely counter intuitive: "How can a rosy breathing body, even one breathing with the help of a machine, be dead?", a person might ask. And she has a point. But never, never, never, never, to breathe again is still what matters behind all the talk and technology.

To begin with, brain-dead people—that is, people whose whole brain, and not just its higher parts, have ceased to function—may subsist quite well in medical facilities and even private homes for decades hooked up to ventilators and sustained by enteral nutrition. The record for "chronic brain death" is now well over 20 years. Brain-dead women have carried their foetuses to term; brain-dead adolescents grow and mature sexually. No death certificate is issued in these cases of the purportedly dead who subsist among us and not in morgues or in the ground. They make ongoing claims on the resources of society if, as occasionally happens, family members insist. (Not many do; most brain-dead people actually die.) They are, in short, not dead except in some metaphorical sense—gone as we knew them perhaps—and are not treated as dead, because we have not yet determined definitively whether they will ever breathe again. (In fact, no properly diagnosed brain-dead person ever has.)[1]

A death certificate with the time and date of death duly noted is not issued until the brain-dead person in question has passed—perhaps failed is the better term—the so called "apnea test:" the breathing test, the never, never, never, never, never test.[2] (The technical term qualification for being really dead in these circumstances is "a positive apnea test".) The brain dead, in short, are not dead until they have demonstrated the "irreversible loss of capacity for spontaneous breathing", which is the modern British medical way of saying what Lear said more poetically.[3] In the United States, the President's Commission for the Study of Ethical Problems in Medicine and Biomedical and Behavioral Research's development of the concept of "Brain Death" puts the King Lear standard more prolixly: permanent inability of an organism to perform its "fundamental work", the main fundamental work being the "drive" exhibited by the whole organism to bring in air.

So-called "brain death" or "whole brain death" is therefore the status of someone who shows no cognitive functions and has taken a series of preliminary examinations that qualify for the only test that in the end matters—the apnea or breathing ever again test. They are rather in the situation of students who have to attend medical school and pass all sorts of examinations in order to gain admission to the one test, which, if passed, qualifies the candidate as a doctor. These qualifying tests for the final apnea test are themselves not new or technologically sophisticated, although some jurisdictions might add electroencephalography (EEG), or other modern procedures to create a visually enduring and rhetorically more demandingly high bar for taking the test that counts. A pronouncement of brain death mobilises century old knowledge to decide whether to subject someone to the King Lear test for the only kind of death there is: never to breathe again.

This is what happens: First, doctors use clinical criteria to rule out other reasons for someone being in a deep and persistent coma rather than whole brain death: abnormally low body temperature; evidence of barbiturate poisoning, for example. Then, they administer a battery of neurological tests, dating back to the nineteenth and early twentieth century, that cumulatively show whether the lower brain, the part that controls breathing, is functional. Each test focuses on one, or several, of the cranial nerves that originate there and control particular reflexes: the oculovestibular reflex is elicited by pouring ice and/or warm (44 °C +) water into the ear and watching for the absence of eye motion which normally works via cranial nerve VIII through Scarpa's vestibular nerve ganglion and the vestibular nuclei in the brainstem (Robert Bárány, a Hungarian Jewish physician, won the Nobel Prize for Medicine in 1914 for this discovery); the gag reflex is tested by inserting a stick into the back of throat; the corneal or blink reflex, is normally produced by touching the cornea, and is absent in those without a functioning brain stem; the photo pupillary reflex is checked by shining a light into the eye and observing whether the pupil contracts; irritating the mucous membranes with ammonia inhalants—smelling salts—would produce the inhalation reflex if the lower brain were intact. There are many more. All bear testimony to the glorious history of nineteenth-century neurology and, cumulatively, to the destruction of the place in the brain that controls breathing as well as so much else. Only when all of these examinations indicate that the brain stem, and hence the whole brain, is indeed dead is a person eligible for the determinative test for death.

The candidate is given a big hit of pure oxygen so that her blood is fully saturated: (10 minutes pre-oxygenation). Then the ventilator is shut off. If she does not breathe within three minutes—in some jurisdictions, five or even eight minutes—she will never never never never breathe again. We know this because the part of the brain that controls breathing is, as the earlier tests had suggested, truly gone. The physiological foundations for this inference go back to the middle of the nineteenth century and were securely settled by the 1920s: in the absence of oxygen, that is, breath, the level of CO_2 in the arterial blood rises; this, in normal circumstances, would raise the acidity of the spinal fluid, which would be registered by the medulla oblongata of the brain stem, which would, in turn, trigger a reflex and produce inspiration. In the United States, a direct measurement of PCO_2 (partial pressure of CO_2) rather than how long the candidate for death has not breathed, is what counts (a value above 60% is considered definitive, but because this level is reached within minutes the two criteria are essentially the same). Now, and only now, after the would-be dead person fails the apnea test, i.e., has a positive apnea test, is it considered as certain as is possible to be— in our macro-physical natural world—that she will never never never never breathe again, and is therefore dead by a standard that was ancient when Shakespeare had King Lear use it. The time of death, to repeat, is recorded not when the patient, already suspected for some time of being brain dead, on the basis of various neurological tests, but when she failed the apnea test and was dead in the old-fashioned way.[4]

In some cases, the body will be relegated to the ordinary fate of dead. In others, it will be hooked up to a ventilator again to keep its organs alive so that they can be harvested for transplantation. Surgeons in keeping with the so-called dead donor rule cannot take a heart and a liver from someone who is alive, nor can they use tissues too long deprived of oxygen. Organ donors are not so much brain dead—they might have been that for a long time—but very recently, i.e., within minutes, really dead, and predictably without breath for ever. The answer to "when is death?" even today is simple: when breath is irrevocably gone.[5]

But of course, it is not quite so easy. Modern technology allows us to widen the scope of "never, never, never, never, never" breath to "never, never, never, never never," something else. "Never to be fully human again" is roughly the standard for what has been called "higher brain death," the death of the brain where memory, reasoning, and consciousness reside. So higher brain death might mean "never to be conscious again" or "never to have what is taken to be the essence of a human being", or "never to recover what had given someone their identity". In these cases, there is no question that the person is dead by any historically defensible category, but rather that she is in such a state that she is already socially and culturally dead and that therefore the living are justified in stopping measures which keep her biologically alive, that is, enteral nutrition and perhaps some assistance in breathing. Of course, we often stop these treatments, assuming for a moment that we consider providing food and air treatments, along with others—antibiotics, vaso suppressors, specific therapies—when death in its old-fashioned sense is taken to be imminent and all interventions hopeless. "Pulling the plug", that is, removing the most critical intervention—the ventilator—is the main way of allowing death to enter. But this is another matter.

The machinery of the intensive care unit has made biological death comport more closely with various conceptions of death as understood culturally even if "really dead" means what it has always meant. If the decision is taken that someone is "never to have the essence of being human again", then the removal of technological life support can translate that decision into biological reality. But technology has done little to alter the rhythms of becoming dead in a broader cultural sense except, perhaps, to expand the lives of the dead among the living.

They speak, as they always have, before and after, at the end of Shakespeare's plays and in many other instances, sometimes as ghosts, but often not. "He being dead yet speaketh," St. Paul said of Abel, slain by his brother Cain. By his faith he speaks from the grave. We are about to read "the work of a dead man," announces the nineteenth-century Brazilian narrator in the prologue of *The Posthumous Memoirs of Brás Cubas* by his compatriot, the novelist Joaquim Maria Machado de Assis. Just to clarify, he wants to make readers aware of the "radical difference between [his] book and the Pentateuch:" Moses waited until the end to speak of the circumstances of his death; our author gives it away at the start. And lest we still miss the point: "I am not exactly a writer who is dead but a dead man who is a writer, for whom the grave was a second cradle". "I expired" he continues, "at two o'clock on Friday afternoon in the month of August, 1869, at my beautiful suburban place in

Catumbi". (Machado de Assis, the other author, died almost four decades later in Rio de Janeiro.)

The very first of the many narrators we encounter in the Nobel prize winner Orhan Pamuk's novel *My Name is Red* similarly begins by telling his readers that he is dead: "I am nothing but a corpse now, a body at the bottom of the well". His head is smashed, his bones are scattered, his mouth filled with blood. He hopes that his wife and children miss him; he isn't sure and thinks, "how dismal it is" that they may have gotten used to his absence. But he hasn't gotten used to being dead: "here, on the other side, one gets the feeling that one's former life persists".

Few of us will not have heard the voices of the dead.

They also continue to work for the living in all sorts of ways. "For us they are not dead", said Hitler of those who were killed in the 1923 Beer Hall Putsch: "this temple is no crypt but an eternal watch. Here they stand for Germany, on guard for our people." They continued, as Caroline Sharples described earlier in this book, Nazi martyrology "to fight for Germany as part of an immortal, spiritual army". These are very old tropes. As early as the eleventh century, King Arthur was also said not be dead at all, but waiting somewhere to return. And the same goes for the twelfth-century Holy Roman Emperor Frederick Barbarossa, who was, and perhaps still is, asleep in his cave in the Kyffhäuser mountains of Thuringia. There, in the 1890s, a gigantic tower, 81 meters high, was built on top of an ancient castle in honour of the dead Kaiser Wilhelm I who could be seen as either the reincarnation of Frederick, or as his successor who presided over the founding, in 1871, of the Second German Empire. Huge statues of both men share the mountain where the dead Barbarossa sleeps. In 1941, the great German assault on the Soviet Union was named Barbarossa. Vladimir Mayakovsky's famous propaganda slogan: "Lenin lived, Lenin lives, Lenin shall live forever" makes the same point.

I would not want to deny that modern medicine, science, and technology have greatly altered the real and imaginative possibilities for blurring the boundaries between life and death and therefore making it more difficult to answer the question "when is death". People who appear fine, if deeply asleep, on ventilators do not seem minutes from death; those declared dead by the old-fashioned criteria of breath still look alive as their organs are harvested. People who are "not there" by many of the criteria we think of as having made them human, and not dead, can be sustained and often breathe on their own. Science fiction, meanwhile, dreams of downloading the whole contents of our brains and offering us a new sort of cyber immortality. Or maybe the answer to "when is death" is never, if our telomeres—the part of human cells that affect how we age—can be manipulated in just the right way.

But the essays in this volume, as well as this Afterword, suggest that little has really changed from very long ago. To be dead is still not to breathe again—ever—and this prognosis is not very hard to make accurately within minutes or hours of its coming to pass. Anyone who has lost someone they love can attest to this. The dead by this standard are gone in almost an instance. Brain death is a distraction from this very basic fact. And the dead still do a great deal for us, individually and collectively, as they have always done. "Becoming really dead," as I wrote elsewhere "takes time".[6] "Never, never, never, never, never" to return to the living as voices, bodies, or ghosts can take years and years, sometimes centuries and millennia.

NOTES

1. At least 12 cases of brain-dead women carrying to term have been reported. For the clinical care of one pregnant dead person see Alan Lane et al., "Maternal Brain Death: Medical, Ethical and Legal Issues", *Intensive Care Medicine*, 30:7 (2004), pp. 1484–1486. Seen from the perspective of transplant surgeons who want to keep brain-dead pregnant women as viable donors see Elizabeth C. Suddaby et al., "Analysis of Organ Donors in the Peripartum Period", *Journal of Transplant Coordination*, 8:1 (1998), pp. 35–39. There are strange exceptions. In 2013 the teenager Jahi McMath suffered irreparable whole brain damage during surgery at Oakland Children's Hospital. After some weeks, she was declared brain dead, took and failed the apnea test—see below—and was issued a death certificate. The hospital refused to keep a legally dead person in one of its beds. McMath's parents refused to accept this verdict and managed to have the body, on a ventilator of course, moved to New Jersey, which allows a religious exemption for brain death. She remains alive there.
2. The word comes from the Greek, *apnous*, breathless (or more specifically, *an absence of*) and a form of the noun *pneúma*, spirit or breath of life; lifeless; that is, dead. See the Liddell-Scott Greek Lexicon, which notes that Galen used the term in its more general and modern sense of "without respiration".
3. See "Criteria for the Diagnosis of Brain Stem Death," *Journal of the Royal College of Physicians of London*, 29:5 (1995), pp. 381–382.

I take my summary of the diagnosis of death following a determination of brain death from the American Academy of Neurology's "Summary of Evidence-Based Guideline for Clinicians" (see http://www.aan.com). See also E.F. Wijdicks et al., "Evidence-based Guideline to Update: Determining Brain Death in Adults", *Neurology*, 74 (2010), pp. 1911–1918. The protocol for children varies slightly. The scores of jurisdictions in the world that accept the diagnosis of brain death each have slightly different protocols.
4. On the history of the physiology of breathing see the posthumous series of lectures by Hans Winterstein, M.D., who more than anyone developed our modern understanding of this subject. "Chemical Control of Pulmonary Ventilation—The Physiology of the Chemoreceptors", *The New England Journal of Medicine*, 255 (1956), pp. 216–223; 227–228; 331–337. International standards and definitions vary widely. See Sam D. Shemie et al., "International Guideline Development for the Determination of Death", *Intensive Care Medicine*, 40:6 (2014), pp. 788–797. For a brief survey of variations across borders which seem no more consistent than they were 10 years earlier see E.F. Wijdicks, "Brain Death Worldwide: Accepted Fact but no Global Consensus in Diagnostic Criteria", *Neurology*, 58:1 (2002), pp. 20–25.
5. It was the needs of organ transplantation programs that motivated the original 1968 Harvard Medical School criteria for brain death and the protocols for determining whether they had been met. See M.L. Tina Stevens, *Bioethics in America: Origins and Cultural Politics* (Baltimore, Maryland and London 2000).
6. Thomas W. Laqueur, "The Deep Time of the Dead", *Social Research*, 78 (2011), p. 802.

AUTHOR BIOGRAPHY

Thomas W. Laqueur is the Helen Fawcett Professor of History at University of California Berkeley. Laqueur's work has been focused on the history of popular religion and literacy; on the history of the body—alive and dead; and on the history of death and memory. His most recent book is *The Work of the Dead: A Cultural History of Mortal Remains* (Princeton University Press, 2016) His current research is on the history of humanitarianism and on dogs in western art.

Open Access This chapter is licensed under the terms of the Creative Commons Attribution 4.0 International License (http://creativecommons.org/licenses/by/4.0/), which permits use, sharing, adaptation, distribution and reproduction in any medium or format, as long as you give appropriate credit to the original author(s) and the source, provide a link to the Creative Commons license and indicate if changes were made.

The images or other third party material in this chapter are included in the chapter's Creative Commons license, unless indicated otherwise in a credit line to the material. If material is not included in the chapter's Creative Commons license and your intended use is not permitted by statutory regulation or exceeds the permitted use, you will need to obtain permission directly from the copyright holder.

Further Reading

Ariès, Philippe. *Western Attitudes Towards Death, from the Middle Ages to the Present*. Patricia M. Ranum trans. (Baltimore, 1975).
Bacigalupo, Félix and Daniela A. Huerta Fernández. "Historical Aspects of the Diagnosis of Death". In Dimitri Novitsky and David K.C. Cooper eds. *The Brain-Dead Organ Donor* (New York, 2013), pp. 7–11.
Bagheri, Alireza. "Individual Choice in the Definition of Death", *Journal of Medical Ethics*, 33 (2007), pp. 146–149.
Baglow, John Sutton. "The Rights of the Corpse", *Mortality*, 12:3 (2007), pp. 223–239.
Bailey, Louis, et al. "Continuing Social Presence of the Dead: Exploring Suicide Bereavement through Online Memorialisation", *Mortality*, 21: 1–2 (2015), pp. 72–86.
Bains, Jatinder. "From Reviving the Living to Raising the Dead; The Making of Cardiac Resuscitation", *Social Science & Medicine*, 47:9 (1998), pp. 1341–1349.
Bartlett, E.T. "Differences between Death and Dying", *Journal of Medical Ethics*, 21:5, pp. 270–276.
Baugh, Bruce. "Death and Temporality in Deleuze and Derrida", *Angelaki: Journal of the Theoretical Humanities*, 5:2 (2000), pp. 73–83.
Bauman, Zygmunt. *Mortality, Immortality and Other Life Strategies* (Cambridge, 1992).
Becker, Ernest. *The Denial of Death* (New York, 1973).
Behlmer, George K. "Grave Doubts: Victorian Medicine, Moral Panic, and the Signs of Death", *The Journal of British Studies*, 42:2 (2003), pp. 206–235.
Behr, John and Conor Cunningham eds. *The Role of Death in Life: A Multidisciplinary Examination of the Relationship between Life and Death* (Eugene, Oregon, 2015).

Belshaw, Christopher. *Annihilation: The Sense and Significance of Death* (London and New York, 2009).
Benatar, David. *Better Never to Have Been: The Harm of Coming into Existence* (Oxford, 2006).
———. Ed. *Life, Death, and Meaning* (Lanham, Maryland, 2004).
Bennett, Gillian and Kate Mary Bennett. "The Presence of the Dead: An Empirical Study", *Mortality*, 5:2 (2000), pp. 139–157.
Bernat, James, Charles Culver, and Bernard Gert. "On the Definition and Criterion of Death", *Annals of Internal Medicine*, 94 (1981), pp. 389–394.
Biggar, Nigel. *Aiming to Kill: The Ethics of Suicide and Euthanasia* (London, 2004).
Bishop, Jeffrey. *The Anticipatory Corpse: Medicine, Power, and the Care of the Dying* (Notre Dame, 2011).
Boonin, David. *A Defense of Abortion* (Cambridge, 2002).
Borgstrom, Erica. "Social Death", *QJM*, 110:1 (2017), pp. 5–7.
———. "Social Death in End-of-life Care Policy", *Contemporary Social Science*, 10:3 (2015), pp. 272–283.
Bradley, Ben, et al. *The Oxford Handbook of the Philosophy of Death* (Oxford, 2013).
Bryant, Clifton D. *Handbook of Death and Dying*, 2 Vols, (Thousand Oaks, California, 2003).
Burkle, Christopher M., et al. "Why Brain Death is Considered Death and Why there Should be No Confusion", *Neurology*, 83:16 (2014), pp. 1464–1469.
Byrne, Paul A., et al. "Brain Death – the Patient, the Physician, and Society". In Michael Potts et al eds. *Beyond Brain Death: The Case Against Brain Based Criteria for Human Death* (New York, 2000), pp. 21–89.
Cann, Candi K. *Virtual Afterlives: Grieving the Dead in the Twenty-First Century* (Lexington, 2014).
Cantor, Norman L. *After We Die: The Life and Times of the Human Cadaver* (Washington D.C., 2010).
Carpentier, Nico and Leen Van Brussel. "On the Contingency of Death: A Discourse-Theoretical Perspective on the Construction of Death", *Critical Discourse Studies*, 9:2 (2012), pp. 99–115.
Caswell, Glenys and Morna O'Connor. "Agency in the Context of Social Death: Dying Alone at Home", *Contemporary Social Science*, 10:3 (2015), pp. 249–261.
Cattorini, Paolo and Massimo Reichlin. "Persistent Vegetative State: A Presumption to Treat", *Theoretical Medicine*, 18 (1997), pp. 263–281.
Cazdyn, Eric. *The Already Dead: The New Time of Politics, Culture, and Illness* (Durham and London, 2012).
Chiong, Winston. "Brain Death without Definitions", *Hastings Center Report*, 35:6 (2005), pp. 20–30.

Cholbi, Michael, ed. *Immortality and the Philosophy of Death* (London and New York, 2016).
Christensen, Dorthe Refslund and Rane Willerslev eds. *Taming Time, Timing Death: Social Technologies and Ritual* (Farnham, 2013).
Clark, David, ed. *The Sociology of Death: Theory, Culture, Practice* (Cambridge, Massachusetts, 1993).
Connolly, Tristanne. *Spectacular Death: Interdisciplinary Perspectives on Mortality and (Un)representability* (Bristol and Chicago, 2011).
Conway, Heather. *The Law and the Dead* (London and New York, 2016).
Craib, Ian. "Fear, Death and Sociology", *Mortality*, 8:3 (2003), pp. 285–295.
Davies, Douglas J. *Mors Brittanica: Lifestyle & Death-Style in Britain Today* (Oxford, 2015).
———. *A Brief History of Death* (Chichester, 2007).
———. *Death, Ritual and Belief: The Rhetoric of Funerary Rites* (London, 2002).
DeVita, Michael. "The Death Watch: Certifying Death Using Cardiac Criteria," *Progress in Transplantation*, 11:1(2001), pp. 58–66.
Dix, Jay and Michael Graham. *Time of Death, Decomposition and Identification: An Atlas* (Boca Raton and London, 2000).
Dubois, James M. "Ethics of Creating and Responding to Doubts about Death Criteria", *The Journal of Medicine & Philosophy*, 35:3 (2010), pp. 365–80.
Ekendahl, Karl. "Death and Other Untimely Events", *Journal of Philosophical Research*, (forthcoming).
Engelhardt, H. Tristram. "Defining Death: A Philosophical Problem for Medicine and Law", *Annual Review of Respiratory Disease*, 112 (1975), pp. 312–324.
Enright, D.J., ed. *The Oxford Book of Death* (Oxford, 2002).
Feit, Neil. "The Time of Death's Misfortune", *Noûs*, 36 (2002), pp. 359–383.
Feldman, Fred. *Confrontations with the Reaper: A Philosophical Study of the Nature and Value of Death* (New York, 1993).
Fernandez, Ingrid. "The Lives of Corpses: Narratives of the Image in American Memorial Photography", *Mortality*, 16:4 (2011), pp. 343–364.
Fischer, John Martin. "Why Immortality Is Not So Bad," *International Journal of Philosophical Studies*, 2 (1994), pp. 257–270.
———. Ed. *The Metaphysics of Death* (Stanford, 1993).
Fustel de Coulanges, Numas Denis. *The Ancient City: A Study on the Religion, Laws and Institutions of Greece and Rome* (Baltimore, 1980).
Gatrell, Vic A. *The Hanging Tree: Execution and the English People, 1770–1868* (New York, 1996).
Gawande, Atul. *Being Mortal: Illness, Medicine, and What Matters in the End* (New York, 2014).
Gennep, Arnold Van. *The Rites of Passage*. Monika B. Vizedom and Gabrielle L. Caffee trans. (London, 1960).

Green, James W. *Beyond the Good Death: The Anthropology of Modern Dying* (Philadelphia, 2008).
Guenther, Lisa. *Solitary Confinement: Social Death and its Afterlives* (Minneapolis, 2013).
Harper, Sheila. "The Social Agency of Dead Bodies", *Mortality*, 15:4 (2010), pp. 308–322.
Harrison, Robert Pogue. *The Dominion of the Dead* (Chicago and London, 2003).
Heidegger, Martin. *Being and Time* (Oxford, 1978).
Himmelman, P. Kenneth. "The Medicinal Body: An Analysis of Medicinal Cannibalism in Europe, 1300–1700", *Dialectical Anthropology*, 22 (1997), pp. 183–203.
Hockey, Jenny, et al eds. *The Matter of Death: Space, Place and Materiality* (Basingstoke, 2010).
Hoffman, Piotr. "Death, Time, History: Division II of Being and Time". In Charles Guignon ed. *The Cambridge Companion to Heidegger* (Cambridge, 1993), pp. 195–214.
Holland, Stephen. "Death as a Biological Category". In Thomas Schramme and Steven Edwards eds. *Handbook of the Philosophy of Medicine* (Dordrecht, 2017), pp. 189–205.
Howarth, Glennys. *Death and Dying: A Sociological Introduction* (Cambridge, 2007).
Howarth, Glennys and Oliver Leaman, eds. *The Encyclopedia of Death and Dying* (New York, 2001).
Hurren, Elizabeth T. *Dissecting the Criminal Corpse: Staging Post-Execution Punishment in Early Modern England* (Basingstoke, 2016).
———. *Dying for Victorian Medicine: English Anatomy and its Trade in the Dead Poor, c.1834-1929* (Basingstoke, 2011).
Jacobe, Stephen. "Diagnosing Death", *Journal of Paediatrics and Child Health*, 51:6 (2015), pp. 573–576.
Jonsson, Annika. "Post-mortem Social Death – Exploring the Absence of the Deceased", *Contemporary Social Science*, 10:3 (2015), pp. 284–295.
Jupp, Peter C. *From Dust to Ashes: Cremation and the British Way of Death* (Basingstoke, 2006).
Jupp, Peter C. and Glennys Howarth, eds. *The Changing Face of Death: Historical Accounts of Death and Disposal* (Basingstoke, 1997).
Kastenbaum, Robert, ed. *Macmillan Encyclopedia of Death and Dying* (New York and London, 2002).
Kaufman, Sharon R. and Lynn M. Morgan. "The Anthropology of the Beginnings and Ends of Life", *Annual Review of Anthropology*, 34 (2005), pp. 317–341.

Kellehear, Allan. "Dying as a Social Relationship: A Sociological Review of Debates on the Determination of Death", *Social Science & Medicine*, 66:7 (2008), pp. 1533–1544.
———— *A Social History of Dying* (Cambridge, 2007).
Klass, Dennis, et al. *Continuing Bonds: New Understandings of Grief* (London, 1997).
Králová, Jana. "What is Social Death?", *Contemporary Social Science*, 10:3 (2015), pp. 235–248.
Kübler-Ross, Elisabeth. *On Death and Dying* (Toronto, 1970).
Lamont, Julian. "A Solution to the Puzzle of When Death Harms Its Victims", *Australasian Journal of Philosophy*, 76 (1998), pp. 198–212.
Laqueur, Thomas W. *The Work of the Dead: A Cultural History of Mortal Remains* (Princeton, New Jersey, 2015).
Lee, Raymond L.M. "The Re-enchantment of Time: Death and Alternative Temporality", *Time & Society*, 18:2-3 (2009), pp. 387–408.
Liechty, Daniel, ed. *Death and Denial: Interdisciplinary Perspectives on the Legacy of Ernest Becker* (Westport, Connecticut, 2002).
Lizza, John P. *Persons, Humanity, and the Definition of Death* (Baltimore, 2006).
Lock, Margaret M. *Twice Dead: Organ Transplants and the Reinvention of Death* (Berkeley and Los Angeles, 2002).
"Death in Technological Time: Locating the End of Meaningful Life", *Medical Anthropology Quarterly*, 10:4 (1996), pp. 575–600.
Luper, Steven. "The Existence of the Dead". In Kasper Lippert-Rasmussen et al eds. *A Companion to Applied Philosophy* (Chichester, 2017), pp. 224–235.
————. *The Philosophy of Death* (Cambridge, 2009).
————. Ed. *The Cambridge Companion to Life and Death* (Cambridge, 2014).
Macdonald, Mary Ellen, et al. "Signs of Life and Signs of Death: Brain Death and other Mixed Messages at the End of Life", *Journal of Child Health Care*, 12:2 (2008), pp. 92–105.
Maddrell, Avril. "Living with the Deceased: Absence, Presence, and Absence-Presence", *Cultural Geographies*, 20:4 (2013), pp. 501–522.
Matteoni, Francesca. "The Criminal Corpse in Pieces", *Mortality*, 21:3 (2016), pp. 198–209.
McCormick, Lisa. "The Agency of Dead Musicians", *Contemporary Social Science*, 10:3 (2015), pp. 323–335.
McCorristine, Shane. *William Corder and the Red Barn Murder: Journeys of the Criminal Body* (Basingstoke, 2014).
McManus, Ruth. *Death in a Global Age* (Basingstoke, 2013).
Moore, Peter. *Where are the Dead? Exploring the Idea of an Embodied Afterlife* (Oxford and New York, 2017).
Nairn, Stuart and Stephen Timmons. "Scientific Uncertainty and the Creation of Resuscitation Guidelines", *Social Theory & Health*, 8:4 (2010), pp. 289–308.

Noys, Benjamin. *The Culture of Death* (Oxford, 2005).
Odile, Frank. "Life After Death: Ethical Issues and Principles of Mental Health Care Professionals in Postmortem Reproduction", *Global Bioethics*, 16:1 (2003), pp. 81–98.
Parkes, Colin Murray, et al eds. *Death and Bereavement Across Cultures* (New York, 1997).
Pernick, Martin S. "Back from the Grave: Recurring Controversies over Defining and Diagnosing Death in History". In Richard M. Zaner ed. *Death: Beyond Whole-Brain Criteria* (Dordrecht and Boston, 1988), pp. 17–74.
Polatinsky, Stefan and Karen Scherzinger. "Dying without Death: Temporality, Writing, and Survival in Maurice Blanchot's *The Instant of My Death* and Don DeLillo's *Falling Man*", *Critique: Studies in Contemporary Fiction*, 54:2 (2013), pp. 124–134.
Prior, Lindsay. "Reflections on the 'Mortal' Body in Late Modernity". In Simon J. Williams et al eds. *Health, Medicine and Society: Key Theories, Future Agendas* (London and New York, 2000), pp. 186–202.
Richardson, Ruth. *Death, Dissection and the Destitute* (Chicago, 2000).
Robson, Jon. "A Time to Die: A Growing Block Account of the Evil of Death", *Philosophia*, 42 (2014), pp. 911–925.
Scarre, Geoffrey. *Death* (London and New York, 2007).
Schor, Esther H. *Bearing the Dead: The British Culture of Mourning from the Enlightenment to Victoria* (Princeton, New Jersey, 1994).
Schumacher, Bernard. *Death and Mortality in Contemporary Philosophy.* Michael J. Miller trans. (Cambridge, 2011).
Seale, Clive. *Constructing Death: The Sociology of Dying and Bereavement* (Cambridge, 1998).
Shewmon, D. Alan. "Brain Death: Can it be Resuscitated?", *Hastings Center Report*, 39:2 (2009), pp. 18–24.
Sinnott-Armstrong, Walter, ed. *Finding Consciousness: The Neuroscience, Ethics, and Law of Severe Brain Damage* (New York and Oxford, 2016).
Steffen, Edith and Adrian Coyle. "Can 'Sense of Presence' Experiences in Bereavement be Conceptualised as Spiritual Phenomena", *Mortality*, 13:3 (2010), pp. 273–291.
Stone, Philip R. "Dark Tourism and the Cadaveric Carnival: Mediating Life and Death Narratives at Gunther von Hagens' Body Worlds", *Current Issues in Tourism*, 14:7 (2011), pp. 685–701.
Strange, Julie-Marie. *Death, Grief and Poverty in Britain, 1870-1914* (Cambridge, 2005).
Sweeting, Helen and Mary Gilhooly. "Dementia and the Phenomenon of Social Death", *Sociology of Health & Illness*, 19:1 (1997), pp. 93–117.
Tarlow, Sarah. "Curious Afterlives: The Enduring Appeal of the Criminal Corpse", *Mortality*, 21:3 (2016), pp. 210–228.

———. *Ritual, Belief and the Dead in Early Modern Britain and Ireland* (Cambridge, 2011).

Tarlow, Sarah and Liv Nilsson Stutz, eds. *The Oxford Handbook of the Archaeology of Death and Burial* (Oxford, 2013).

Taylor, James Stacey. "Death, Posthumous Harm, and Bioethics", *Journal of Medical Ethics*, 40 (2014), pp. 636–637.

Valentine, Christine. *Bereavement Narratives: Continuing Bonds in the Twenty-first Century* (London and New York, 2008).

Van Brussel, Leen. *Social Construction of Death: Interdisciplinary Perspectives* (Basingstoke, 2014).

Verdery, Katherine. *The Political Lives of Dead Bodies: Reburial and Postsocialist Change* (New York, 1999).

Vidal, Fernando. "Brainhood, Anthropological Figure of Modernity", *History of the Human Sciences*, 22:1 (2009), pp. 5–36.

Walter, Tony. "The Sociology of Death", *Sociology Compass*, 2 (2008), pp. 317–336.

———. "Plastination for Display: A New Way to Dispose of the Dead", *Journal of the Royal Anthropological Institute*, 10:3 (2004), pp. 603–627.

Whetstine, Leslie M. "The History of the Definition(s) of Death: From the 18th Century to the 20th Century". In David W. Crippen ed. *End-of-Life Communication in the ICU* (New York, 2008), pp. 65–78.

White, Carol J. *Time and Death: Heidegger's Analysis of Finitude*. Mark Ralkowski ed. (Aldershot, 2005).

Wijdicks, Eelco F.M. "Determining Brain Death", *Continuum: Lifelong Learning in Neurology*, 21:5 (2015), pp. 1411–1424.

Wójcicka, Natalia. "The Living Dead: The Uncanny and Nineteenth-Century Moral Panic over Premature Burial", *Styles of Communication*, 2:1 (2010), pp. 176–186.

Yanke, Greg, et al. "When Brain Death Belies Belief", *Journal of Religion and Health*, 55:6 (2016), pp. 2199–2213.

Youngner, Stuart J. "Philosophical Debates about the Definition of Death: Who Cares?", *The Journal of Medicine and Philosophy*, 26:5 (2001), pp. 527–537.

Youngner, Stuart J. et al eds. *The Definition of Death: Contemporary Controversies* (Baltimore, 1999).

Zivkovic, Tanya. "Returning from the Dead: Contested Continuities in Tibetan Buddhism", *Mortality*, 18:1 (2013), pp. 17–29.

Index

A
Albinus, Bernhard Siegfried, 62
Alzheimer's disease, 109
Apnea test, 14, 147–149
Ariès, Phillipe, 2, 133, 142
Armitage-Moore, Maisie, 43
Austin, J.L, 18
Axmann, Artur, 96

B
Bárány, Robert, 148
Barbarossa, Frederick, 151
Battle of Carbisdale, 36
Behlmer, George, 70
Bernadotte, Folke, 91
Bezymenski, Lev, 88
Bierce, Ambrose, 137
Bland, Anthony, 121, 122, 125, 126
Blaschke, Hugo, 94, 95
Boleyn, Anne, 28
Bonaparte, Napoléon, 59
Borneman, John, 90
Bourget, Paul, 141, 142
Brain death, 13, 14, 120–122, 127, 128, 147, 148, 150, 152

Braun, Eva, 88, 91, 93, 94, 96, 97
British Embalmers' Society, 74, 77
British Institute of Embalmers, 79
British Undertakers' Association, 74, 77
Browne, Thomas, 2
Buchan, John, 35
Buckhout, O.K, 79
Buddhism, 114
Burbage, Richard, 26
Burggraeve, Adolphe, 51, 55–57, 60, 62, 63
Burial, 1, 4, 7, 10, 12, 13, 38–40, 49, 53, 71–74, 76–78, 80, 81, 105, 107, 113, 114, 127
Burns, Robert, 108

C
Campbell, Archibald, 1st Marquis of Argyll, 35, 37
Camus, Albert, 140
Capote, Truman, 139, 140
Chambers, Robert, 35
Chapman, Hugh, 78
Charles I, 33, 35, 36, 38, 39, 41

Charles II, 34, 36, 39
Christianity, 51, 104
Churchill, Winston, 93
Church of England Funeral and Mourning Reform Association, 75
Clutter, Herb, 139
Conrad, Brian, 93
Conservation, 50–53, 56, 58–61, 63
Cottridge, Albert, 77, 79
Coulanges, Fustel de, 2
Crace, Jim, 17, 18
Cremation, 1, 12, 70, 72, 74–77, 80, 81, 94, 107, 114
Cremation Society of England, 72
Cromwell, Oliver, 40, 44

D
Decomposition, 7, 10, 17, 50, 71, 74, 77, 81, 146
Descartes, René, 123
Dickens, Charles, 137
Dismemberment, 42
Dissection, 2, 56, 59
Donald, Merlin, 128
Dönitz, Karl, 90–93, 97, 98
Donne, John, 23
Down, Langdon, 70
Drummond, David, 78
Durham, Alexander, 40

E
East India Company, 42
Echtmann, Fritz, 94, 95
Edgeworth, Maria, 8
Eichmann, Adolf, 98
Eisenhower, Dwight, 92
Elser, Georg, 89
Embalmment, 54, 58, 59, 74
Embodiment, 108, 127, 129
Erskine, Elizabeth, Lady Napier, 41

Eucharist, 111
European Convention of Human Rights, 124
Excommunication, 11, 37
Execution, 2, 11, 13, 36–38, 40, 42–44, 136–139, 141

F
Fellows, Alfred, 78
First Bishop's War, 35
First Marquis of Montrose Society, 44
Frazer, James, 107

G
Gannal, Jean-Nicholas, 58–60, 63
Gibbet, 2
Giesing, Erwin, 93
Goebbels, Joseph, 93
Goebbels, Magda, 93
Goff, Robert Lionel Archibald, 122
Gorer, Geoffrey, 109
Graham, James, 1st Marquis of Montrose, 11, 33, 37, 39, 43
Graham, John, 41
Graham, Thomas, 41
Greenwood, George, 78
Grey, Henry, 1st Duke of Suffolk, 139
Grief, 13, 54, 107–109

H
Haden, Francis Seymour, 75
Hadwen, Walter, 75, 80
Hagens, Gunther von, 11
Harris, John, 124, 127
Hart, Ernest, 26, 72
Heidegger, Martin, 2, 134
Hendriksen, Marieke, 52
Hertz, Robert, 108
Heusemann, Käthe, 94, 95

Hickock, Richard, 139, 140
Himmler, Heinrich, 91–93
Hitler, Adolf, 12, 87–98, 151
Hoffmann, Leonard Hubert, 122
Huntly, George, 2nd Marquis of Gordon, 40
Hurry, James, 78, 80

I
Islam, 105

J
Johnson, Alexander, 42, 136
Johnson, Samuel, 42
Judaism, 106

K
Kant, Immanuel, 123
Karma, 106, 108
Karnau, Hermann, 94, 96
Keening, 6, 10
Knowles, 43
Koan, 114, 115
Krüger, Elsa, 95
Kübler-Ross, Elisabeth, 2
Kyd, Thomas, 19

L
Lanzmann, Claude, 136
Lawrence, Frederick, 75
Lenin, Vladimir I, 151
Litvinov, 93
Lizza, John P, 127
Locke, John, 123
London Association for the Prevention of Premature Burial, 69, 75
Lozovski, 93
Lucan, Lord, 12

M
MacDonald, Catriona, 36
Machado de Assis, Joaquim Maria, 150, 151
Mackenzie, George, 35
Mansfeld, Erich, 96
Mantel, Hilary, 28
Marchetti, Daniela, 88
Maron, Karl, 96
Marris, Peter, 109
Mayakovsky, Vladimir, 151
McLeod, Neil, 36
McMahan, Jeff, 123
Melville, Herman, 137
Meulewaeter, Edouard, 55
Michiels, Joseph, 61
Mill, John Stuart, 140
Mills, Halford Lupton, 75
Moorhouse, Roger, 95
Morkill, Alan Greenwood, 41
Morkill, J.W, 41
Mormonism, 104, 106
Müller, Willi Otto, 96
Mussolini, Benito, 90, 91

N
National Association of Funeral Directors, 81
National Council for the Disposition of the Dead, 80
National Health Service, 80

O
O'Collun, Patrick, 28
Operation Valkyrie, 89
Organ donation, 108
Orwell, George, 138, 139

P

Palmer, Charles James, 70
Palmer, Katherine Millicent, 70
Pamuk, Orhan, 151
Pernick, Martin, 127
Personhood, 49, 53, 62, 109, 123, 126–129
Peterson, Kaara, 21
Petrova, Ada, 88
Pickering, 41
Pickersgill, Jeanette Caroline, 70
Plato, 137
Platter, Thomas, 21, 26
Plunkett, Oliver, 44
Poe, Edgar Allan, 70
Poelman, Charles, 60
Pogonyi, J.F., 94
Post-mortem punishment, 38
Premature burial, 12, 69, 70, 72, 75–81
President's Commission for the Study of Ethical Problems in Medicine and Biomedical and Behavioral Research, 147
Presley, Elvis, 12
Puckle, Bertram, 73
Purgatorian Society, 7

R

Rathlin Island, 5
Reeves, C., 41
Relics, 34, 41, 44, 51
Revivification, 21
Roemer, Gijsbert M. van de, 52
Russ, William, 71
Ruysch, Frederik, 50, 51

S

Scott, Walter, 36
Serov, Ivan, 94
Seymour, Henry, 141
Shakespeare, William
  Hamlet, 19, 22, 24–27, 29
  King Lear, 22, 27, 29, 149
  Othello, 11, 20, 26
  Romeo and Juliet, 19, 20
Sherry, Henry, 77
Shildrick, Margrit, 123
Sidney Sussex College, Cambridge, 46
Singer, Peter, 124
Smith, Perry, 139
Socrates, 137
Soul, 6, 7, 25, 50, 54, 60, 105, 106, 112
Spiritualism, 25, 34, 35, 44, 88, 110, 136, 151
Stalin, Joseph, 90, 94, 95
Stillbirth, 77, 79, 104
Strange, Julie-Marie, 54
Suicide, 12, 20, 25, 26, 73, 87, 88, 94, 98, 113
Survival, 12, 87–89, 97, 98, 133
Szechi, Daniel, 38

T

Tarlow, Sarah, 50
Tebb, William, 75
Teptzov, N.V., 94
Thompson, Henry, 72, 76
Thoresby, Ralph, 41
Tolstoy, Leo, 135, 143
Traudl, Junge, 95
Treaty of Berwick, 35
Trevor-Roper, Hugh, 88, 95
Troyer, John, 62

V

Valentine, Christine, 111
Vinogradov, V., 94
Vollum, Edward, 75

W

Waldo, Frederick, 76
Watson, Peter, 88
Webster, John, 20
Weekend at Bernie's, 2, 3
Weidmann, Eugen, 136
Wells, Thomas Spencer, 72
Wheatley-Crowe, H. Stuart, 43
Wilhelm I, 151

Willis, Thomas, 146
Wilson, Emily R., 28
Wishart, George, 37
Woodrow, Robert, 35

Z

Zhukov, Georgy, 94

Open Access This book is licensed under the terms of the Creative Commons Attribution 4.0 International License (http://creativecommons.org/licenses/by/4.0/), which permits use, sharing, adaptation, distribution and reproduction in any medium or format, as long as you give appropriate credit to the original author(s) and the source, provide a link to the Creative Commons license and indicate if changes were made.

The images or other third party material in this book are included in the book's Creative Commons license, unless indicated otherwise in a credit line to the material. If material is not included in the book's Creative Commons license and your intended use is not permitted by statutory regulation or exceeds the permitted use, you will need to obtain permission directly from the copyright holder.

The manufacturer's authorised representative in the EU is Springer Nature Customer Service Centre GmbH, Europaplatz 3, 69115 Heidelberg, Germany. If you have any concerns regarding our products, please contact ProductSafety@springernature.com

Printed and bound by CPI Group (UK) Ltd, Croydon, CR0 4YY

23/03/2026

02076401-0011